[illegible]
Wash. D.C.

Bought Sept 1978

THE UNCERTAIN PROMISE

THE UNCERTAIN PROMISE

VALUE CONFLICTS IN TECHNOLOGY TRANSFER

DENIS GOULET

Published by
IDOC/North America, New York
in cooperation with the
Overseas Development Council, Washington D.C.

The Uncertain Promise:
Value Conflicts in Technology Transfer

Published by IDOC/North America, Inc.
145 East 49th Street, New York, New York 10017
in cooperation with the
Overseas Development Council
1717 Massachusetts Avenue, N.W., Washington, D.C. 20036

ISBN 0-89021-044-6 *cloth*
ISBN 0-89021-045-4 *paper*

Library of Congress catalog card number
77-81314

Printed in the U.S.A.

Cover design by Terrence J. Gaughan
Typesetting by Prime Publishing Services, Inc., New York, New York
Printed by Edwards Brothers, Inc., Ann Arbor, Michigan

For Reynaldo, Isa, and their "tribe"

Contents

Books by Denis Goulet

ETHICS OF DEVELOPMENT

THE MYTH OF AID: THE HIDDEN AGENDA OF THE DEVELOPMENT REPORTS
(with Michael Hudson)

A NEW MORAL ORDER: DEVELOPMENT ETHICS AND LIBERATION THEOLOGY

THE CRUEL CHOICE: A NEW CONCEPT IN THE THEORY OF DEVELOPMENT

Preface

This study of value conflicts in technology transfers to the Third World is based on documents reviewed and field research conducted from September 1972 to December 1975. During this period I interviewed scholars, government and international agency officials, patent lawyers, licensing agents, corporate buyers and sellers of technology, consultants, engineers, directors of industrial research laboratories, bankers, peasants, and factory workers. Both formal questionnaires and open-ended interviews were designed to elicit, from respondents engaged in technology activities, critical statements as to their values and priorities. One technique proved especially fruitful: the "match-up" interview in which partners to technological contracts were queried, separately and in their respective work sites, regarding their dealings across national boundaries.

Data-gathering centered on the United States, Argentina, Brazil, Chile, and Peru. The following countries served as secondary research sites: Canada, France, Great Britain, Algeria, Colombia, Venezuela, and Bolivia.

Countless individuals opened the doors of their minds and work places to me. Although I cannot here acknowledge them all by name, I thank them heartily. Several persons were prodigal in meeting my research needs: Hector Font (Venezuela), Enrique Iglesias (Chile), Antoine Kher (France), Horacio Rodriguez Larreta, and André van Dam (Argentina), and William Krebs and Eugene Moore (USA). From Sheldon Gellar, Thomas Fox, Pierre Gonod, John Sewell, John Lewis, Jack Behrman, Morris Morris, and Harald Malmgren I received valuable suggestions for manuscript revision.

The following institutions provided a stimulating work base: Center for the Study of Development and Social Change, Organization of American States, and Overseas Development Council. I am particularly thankful to the following persons at Overseas Development Council: James Grant, John Sewell, Guy Erb, Michael O'Hare, and Valeriana Kallab.

This project would not have been possible without generous financial assistance from the Center for Studies of Metropolitan

Problems (Metro Center), National Institute of Mental Health, under Research Grant No. 5 FO3 MH54828-02 MP. My gratitude extends to all at Metro Center, especially to Richard P. Wakefield, for his unfailing and imaginative help.

A grant from the Inter-American Foundation covered the services of a research assistant during the last fourteen months of work. My thanks, therefore, go to William Dyall and Csanad Toth for their "tailor-made" assistance.

Mary Lesser, Debbie Carlson, Georgia Shelton, and Phyllis Jansen served me well as research assistants at successive stages of work, and Pat Gaughan at IDOC/North America (assisted by Paul Hallock) has helped greatly with editorial suggestions.

My intellectual debts to colleagues are numerous, but I must single out Francisco Suarez and Louis Xhignesse who have been, for many years, at once catalysts of new ideas and peerless friends.

The most important acknowledgement comes last. As a "philosopher of development" whose life is replete with value conflicts and overlapping loyalties, I have been aided beyond measure by my wife, Ana Maria Reynaldo. She is my most demanding critic as well as the main inducement I have for keeping my eyes fixed on essential values. It is, accordingly, a singular pleasure and honor for me to dedicate this book to her parents and to her numerous "tribe" dispersed throughout Brazil.

No one mentioned here can be held responsible for defects or errors found in this work. For these creatures of my own making, I assume full responsibility.

Introduction

Introduction

Is modern technology the key to successful development? Will technology "deliver" on its promise to bring development to the Third World? Can technology be "transferred" from one cultural setting to another in ways which are more beneficial than destructive? And how do policies for becoming technologically "modern" relate to broader development goals in diverse nations? These questions lie at the heart of the present study. Technology is portrayed herein as a "two-edged sword," simultaneously the bearer and destroyer of values. Yet, although it originates in "developed" societies, modern technology circulates rapidly in the world through a variety of transfer channels. Voluminous writings on "technology transfer" have now made their way into the literature on development.[1] *The present work focuses on value conflicts in transfers of technology from rich to poor countries.* Conflicts arise at several levels: (a) competing interests of buyers and sellers of technology; (b) tensions between overall development goals and the impact of imported technology on poor countries; and (c) general questions as to the possibility of harnessing technology in any society to such humane ends as a satisfying scale of operations, ecological soundness, and the just allocation of resources.

Technology is a vast domain touching all sectors of human activity. A question arises, therefore, as to which technology transfers are included and which are excluded from the purview of this work. Because some choice bearing on manageability of data had to be made, I have concentrated on industrial technologies used in production, with only secondary attention paid to agricultural and communication technologies. Moreover, while no student of development ignores the impact of military technology transfers, these are not discussed here because proper study of them requires special access to data which is lacking to me. Nevertheless, a focused examination of industrial technology sheds much light on the dynamics of transfer and the implications of uneven bargaining positions (and no one can gainsay the importance or frequency of such transfers in processes of development). I have also confined this study to transfers conducted by private enterprises, to the exclusion of initiatives undertaken by governments and their agencies.

A word of caution is in order because of the loose terminology in vogue on the subject. Many technology "transfers" conducted by governmental or international agencies are not transfers in the strict sense but rather training efforts designed to improve skill levels of actors in technological arenas. More strictly viewed, technology transfers relate to the circulation of know-how which is directly applied to the production of goods, the provision of services, and the formulation of decisions (as to site, engineering design, scale, etcetera) affecting these.

The present book centers on transactions between the United States and Latin America, but even within that continent, comprehensive coverage has not been attempted. Moreover, because broad differences between large and small nations, structurally complex societies and those less so, and those possessing different technical capacities are mirrored in the larger Third World (including Asia, Africa, and the Arab countries), the value of generalizations drawn from this study will be limited. Nevertheless, value conflicts in industry do illustrate larger patterns of technological dependency, tensions between efficiency and social goals, and the crucial importance of overall incentive systems in any society. An important distinction must also be made between those value conflicts which pit Third World *nations* against rich-country exporters of technology and those which weigh on poor *populations* whenever technological decisions are monopolized by a specialized elite, whether foreign or indigenous. My analysis is situated primarily at the level of relations conducted across national boundaries. Nonetheless, tensions between national decision-makers and nonexpert majorities hover constantly in the background, especially when criteria for technology policy are discussed and incompatibilities surface between technological efficiency and broader goals such as social justice or job-creation. To oscillate from one pole to another is not fully avoidable, inasmuch as strong correlations exist between Third World countries' attempts as nations to improve their relative bargaining stance in international technology arenas and their domestic efforts to remove technological power from the hands of a few specialists. Clearly there can be no "technology for the people" if control is vested in foreign corporations. On the other hand, shifting control from foreigners to national technicians, politicians, or entrepreneurs cannot, of itself, guarantee a policy which benefits the masses.

One major premise of this book is the existence of the vital nexus which links the value content of modern technology to basic development strategies and to technology policies adopted in less-developed countries. Given this linkage, I examine value conflicts in a mode which combines philosophical and empirical inquiry. The philosophical character of the book is most apparent in Part One, which analyzes technology as a universe having its own values and dynamisms, and in the Conclusion, where basic questions of freedom and the quality of civilization are raised. Empirical dimensions dominate in Parts Two and Three, which deal, respectively, with the impact of technology transfer on development and with technology policies for Third World countries.

My earlier books center on ethical and value questions posed by development theory, planning, and practice.[2] As a philosopher of development, I adhere to the view that the complex realities of "development," in their dual nature as social change processes of a special kind and as an array of competing images of the good life and the just society, are best understood by focusing on the value conflicts they pose. The present book builds, therefore, on the conception of development presented in earlier works. It may prove helpful, at this point, to state what this conception is by citing from a previous book.

> Development ethics borrows freely from the work of economists, political scientists, sociologists, planners, and spokesmen for other disciplines. Although each discipline supplies its own definition of development, ethics places all definitions in a broad framework wherein development means, ultimately, the quality of life and the progress of societies toward values capable of expression in various cultures. Along with the late L.J. Lebret, I view development as a complex series of interrelated change processes, abrupt and gradual, by which a population and all its components move away from patterns of life perceived in some significant ways as "less human" toward alternative patterns perceived as "more human." *How* development is gained is no less important than *what* benefits are obtained at the end of the development road. In the process new solidarities, extending to the entire world, must be created.[3] Moreover, cultural and ecological diversity must be nurtured. Finally, esteem and freedom for all individuals and societies must be optimized. Although development can be studied as an economic, political, educational, or social phenomenon, its ultimate goals are those of existence itself: to provide all men with the opportunity to live full human lives. Thus understood, development is the ascent of all men and societies in their total humanity.[4]

The present work moves beyond these general issues and formulates ethical strategies in one specific arena of development *decision-making*, namely, technology policy. Hence, although acquaintance with the earlier works is helpful to the reader, it is not indispensable. What is necessary, however, is an understanding of how I am defining and using three key terms: *technology, transfer,* and *values.*

Definitions

Technology may be defined as the systematic application of collective human rationality to the solution of problems by asserting control over nature and over human processes of all kinds.[5] Technology is normally the fruit of systematic research which is disciplined and cumulative, not merely accidental or serendipitous. Moreover, it is not mere intellectual speculation or theoretical modeling but rather knowledge *applicable* to practical problems. And further, this systematically applied human reason must operate in a *collective* societal context, so that a practical invention which originates in a solitary mind does not qualify as technology unless it is expressed in a tool, process, or object which can be used by others. Technological activity, then, aims at expanding and improving the ability of human beings *to control* the natural and social forces which surround them.

To analyze value conflicts arising from applying technology to development in poor countries is a major aim of this book. Because most technology originates outside these countries, one must define technology *transfers.*

For British economist Charles Cooper *technology transfer* is "the transfer or exchange from advanced to developing countries of the elements of technical know-how which are normally required in setting up and operating new production facilities and which are normally in very short supply or totally absent in developing economies."[6] This definition is not comprehensive enough, however, because there are also many technologies which do not relate directly to the operation of product-producing or -processing facilities. Thus, the concept of transfers must also embrace the circulation of know-how used to conduct feasibility or marketing studies and to manage varied services (transportation, distribution, etcetera). Also to be included are mastery of the criteria for evaluating and choosing from among numerous technological alternatives and the kinds of specialized knowledge needed, for example, to engineer designs or construct plants. All these are technologies which are transferable from one institutional setting to another. But the term *transfer* as used in this work refers to the circulation of know-how *across* national boundaries; excluded from consideration are transfers of technology from one sector to another within national societies—the application of findings obtained from space research, let us say, to the housing industry or to the manufacture of electrical appliances.

A third term remains to be clarified, namely, *values*. Although specialized definitions abound, in common discourse *values* refers to attitudes, preferences, styles of life, normative frameworks, symbolic universes, belief systems, and networks of meaning which human beings give to life. Social scientists, legal scholars, philosophers, poets, theologians, and historians all experience great difficulty in defining *values* with precision and realism. For working purposes, however, I take a value to be any object or representation which can be perceived by a subject as habitually worthy of desire.[7] All agree that value entails "oughtness," a note captured in the foregoing definition by the element of "worthiness." And whether it is rooted in object or subject, value is only operative when perceived; hence the concept of perceptibility is essential to a definition. The word *desire* suggests that value is not a purely cognitive category but can involve will, emotion, and passion. Because images and representations can move the will and mobilize energies as readily as real objects, they too must be treated as values. Values, in short, are all goods—real or imagined—which stir people's desires, command their allegiances, and move them to act.

This study in value conflicts examines competing visions of benefits sought in technological exchanges. Many conflicts are traceable to technology itself, not merely to the mechanisms by which it is "transferred" from one national setting to another. Accordingly, before subjecting the transfer process to scrutiny, I have analyzed technology itself as the simultaneous bearer and destroyer of values. The relation between technology and development must first be elucidated, however. And in order to do that, a prior question must be asked: Is technology the key to successful development?

Technology: The Key to Development?

Technology affects development on four counts: It is a major resource for creating new wealth; it is an instrument allowing its owners to exercise social control in various forms; it decisively affects modes of decision-making; and it relates directly to patterns of alienation characteristic of affluent societies.

Technology as Resource

Those who avidly seek development await such benefits as improved material living standards and new wealth through greater production and productivity. Most development agents assign an important role to technology in reaching these objectives. Indeed, if suitably chosen and properly applied, technology can add greatly to a society's pool of resources. Hence, U Thant, upon inaugurating the United Nations Second Development Decade in 1970, called technology the single most important resource needed to create other resources. And one group of experts has written that "scientific and technical information is the lifeblood of progress in a technically advanced society. It is just as much a natural resource as the stock of laboratories and trained personnel."[8] Technology as a resource must obviously be distinguished from other types of resources, particularly "natural" resources which exist in different states of exploitability.[9]

Many resources, in turn, are not natural but human creations—capital, machinery, infrastructures of every sort. And human skills themselves range from strong muscles for pushing wheelbarrows to elaborate forms of knowledge needed to program computers. Technology is merely one special kind of human skill: the know-how derived from scientific knowledge and incorporated in some object, process, or activity. Here lies the basis for distinguishing among technologies which are product-embodied, process-embodied, or person-embodied.[10] This division is obviously somewhat artificial, because all technology is person-embodied. Nevertheless, certain technologies are incorporated in concrete tools or products, whereas others are not. For example, many technologies are embodied in such simple tools or products as hammers, screwdrivers, stamping machines, and ball bearings. Process-embodied technologies, in turn, are incorporated in plans, formulas, blueprints, and directions for processing materials into finished products. Process technologies presuppose detailed knowledge of the properties of chemicals or physical elements and cumulative experience born of trial and error as to the best sequence to follow in heating, melting, or blending. These technologies are sensitive to slight differences in temperature, density, speed of flow, or pressure. The third category, person-embodied or decisional technology, embraces the practical knowledge used by planners, designers, managers, technicians, and engineers in analyzing bodies of information to determine what practical consequences may be drawn from them. The preferred tools of experts initiated in the arts of diagnosis, decision, and management are models, abstract symbols, and other relatively intangible instruments.

Varying degrees of *tangibility* in resources have proved important over time. For centuries most economic resources consisted of tangible goods supplied by nature: fertile lands, plentiful game, abundant ores, lush timber. As economic life grew more complex, however, humanly created resources took on greater relative importance. Labor skills, communications networks, symbols which conferred meaning to tasks, and social-incentive systems all became "factors" of production. With the advent of capitalist industrialism, still other "resources" gained salience: entrepreneurial skills, access to diversified markets, and negotiating talents. Predictably, mechanical "invention" proved greatly advantageous at first to those who could harness it "economically" to productive processes. Eventually, however, the very process of generating invention came to be subsidized, mainly by governments seeking competitive advantages in weaponry. Thus the creation of technology or the "invention of invention" became an economic activity, responsive to pressures of demand and supply. Laboratories were built as factories for producing technology, which came to be perceived as a special capital good with an exceptionally high multiplier effect.

Technology is a key resource because those who possess it can be presumed capable of possessing finance capital and other resources as well. Thanks to technology, one can create substitutes for many "natural" resources. Moreover, when technology is applied to the existing pool of other resources, quantum leaps in productivity can be expected. Conversely, without technology, even an abundance of other "factors" of production augurs ill for economic success.

But technology never exists in a social vacuum; it is owned by identifiable interest groups who may use it as an instrument of social control.

Technology as Instrument of Social Control

Transnational corporations have long used their special mastery of technology to gain for themselves not only economic advantages but cultural and political influence as well. Military forces, in turn, have parlayed their technological superiority over other institutions into control over the direction of social change in many countries. Professionals in medicine, engineering, education, architecture, and industrial design habitually invoke their technological expertise to define limits within which the utilizers of their services are to have their needs met. In these domains social control takes the form of exercising a monopoly in the diagnosis or prepackaging of professional services, a subtle but efficacious means of controlling demand.[11]

The exercise of social control by technological elites is greatly facilitated by the arcane language and symbolism they employ. If knowledge is power, then esoteric knowledge is, by definition, inaccessible power. Technological knowledge confers upon those who possess it the power to define problems, to delimit alternative solutions, and to influence outcomes.

Many Third World leaders trace their subservience in vital decision-making arenas to their technological weakness. Consequently, they seek modern technology not only for economic reasons but also to reduce their vulnerability to control by technologically advanced nations. Technological mastery is indeed a passport to decision-making power.

Technology and Decision-Making

Modern decision-making takes on the values, biases, and dynamisms associated with technology. One must ask whether technology inherently fosters elitist modes of decision-making. The answer to this question will reveal what degree of participation one deems possible in planning. That is, if one believes that sound decisions can be reached only after analysis by "experts," it is futile to preach democratic participation. If, on the contrary, one concludes that technology itself makes it possible for decision-makers to "go to the people" to learn their wishes, then no excuse exists for elites to restrict access to their special information; to do so constitutes a breach of the developmental promise. Some contend that today's problems are so complex that only specialists know how to define problems and list possible solutions. Others retort that this very complexity condemns elitism to failure and makes it necessary for the presumed beneficiaries of technology to express their values and aspirations before choices are made. This serious question is cited here merely to suggest how wide is the scope of technology's impact.

"Organic" and "mechanistic" approaches to decision-making contrast sharply. Conventional decision-makers are fond of briefings in which their advisers "lay out the range of options." Yet this practice systematically excludes holistic considerations not amenable to encapsulation in briefings or catalogues of options. Furthermore, in such conflictual domains as politics, personal life, and corporate existence, the emotions, fears, and perceived threat of actors are crucial variables affecting decisions; yet they are eliminated from mechanistic processes.[12] Organic debate and interaction, on the other hand, are the normal preludes to sound compromises and decisions. The question is whether technology's proper dynamism is compatible with these modal values in decision-making.

Feedback operations are essential. One may plausibly argue that to structure feedback is merely to assure that any participation elicited will be a mere "reaction" to what is proposed. To *propose* is thus, in effect, to *impose*, inasmuch as those who plan initial arrangements do not provide for a *feed-in* at early moments of problem-definition. Feedback prevents nonexperts from gaining access to essential parameters of the decision process *before* these are congealed. Some reject this portrayal and insist that initial formulations of problems and alternative solutions are themselves the product of some prior exchange in which a first round of feedback leads to a second round of feed-in, and so on. Whatever be one's position, one stark conclusion emerges: The technological feasibility of circulating information and counterinformation conditions decisional modes.

Technology tends to privilege decisions based on will. Thanks to technology, new products, processes, and systems can be willed into existence. But technology can neither generate, nor perhaps even tolerate, the play of several qualities essential to good decisions: compassion, a concern for justice, intuition, or empathy. As decisional procedures come to depend increasingly on technology, the danger arises that decision-makers with a strong will to power will gain alarming advantages over peers whose primary allegiance is to justice or compassion. That this danger is neither illusory nor remote is illustrated by recent trends in medical and genetic technology as applied to social engineering.[13]

Technology and Alienation in Abundance

Aristotle warned long ago that "the amount of property which is needed for a good life is not unlimited."[14] He qualified Socrates' dictum that "a man should have so much property as will enable him to live temperately, which is only a way of saying 'to live well.'" For Aristotle, the good life includes both temperance and liberality—the relative emancipation from want. The two must be conjoined, for otherwise "liberality will combine with luxury; temperance will be associated with toil."[15] This evocation of luxury suggests an interesting line of reasoning adumbrated in *Luxury and Capitalism* by Werner Sombart, one of whose conclusions is that capitalism was made dynamic by the quest for luxury. Veblen's *The Theory of the Leisure Class* is largely a variation on the same theme. Significantly, both works claim that capitalism became truly dynamic when it harnessed technology to production. At first, technology allowed those owning capital goods to enjoy luxury, of which leisure is merely one socially conspicuous form. Over time, however, technology made luxury and its surrogates available to mass consumers. The dissemination of comforts and ornaments had a deep psychological effect. In earlier times only the idle rich were struck by that special *ennui* which affects those who have savored every luxury and found them all equally unrewarding. Now, however, it became possible for the masses to experience the boredom and alienation that come from striving, obtaining, and remaining unsatisfied. So long as the majority in any society lacked enough goods to be sated—or even plausibly to imagine that their happiness lay in the possession of goods—they were sheltered from this particular form of suffering. The meaningfulness of their lives could not realistically be thought to reside in superfluous material satisfactions. The only material satisfactions they could realistically enjoy came closer to primary needs—biological, psychical, or spiritual. The opposite of alienation, however, is not satisfaction but meaningful living. And here centers the revolutionary impact wrought on the human psyche by technology: It has stripped societies and their members of their sources of meaning. Pretechnological societies derived their meaning from synthesis, whereas technology has destroyed the basis for any synthesis other than its own, which is dry and sterile. Technology need not have wrought this destruction, but historically it has. One glimpses the cultural drama entailed here in such development novels as Cheikh Hamidou Kane's *Ambiguous Adventure*[16] or Chinua Achebe's *Things Fall Apart.*[17] The same phenomenon has been eloquently "filmed" in words by sociologist Benjamin Barber in his recent study of technology and Swiss cantons.[18] Earlier literary portraits are found in such nihilistic novels of the nineteenth century as Turgenev's *Fathers and Sons.* Here the iconoclastic hero, Bazarov, enthusiastically tears down all conventional values because he holds in his hands a new substitute, science. Finally he makes a long train journey, carrying nothing but a suitcase filled with straw. When his baggage is discovered by those he formerly taunted, Bazarov embarrassingly replies that he cannot go through life carrying an empty suitcase. The empty suitcase—a fit paradigm of the spread of alienation in "developed" countries!

Galloping alienation is the price exacted of societies which pursue technological success competitively. Thus pursued, technology reduces the totality of human meaning to those of its elements which are amenable to problem-solving. Widespread alienation is an eloquent sign that the meaning of life hitherto supplied by cultural belief systems cannot be replaced by technology. Psychologist Erich Fromm points to the equally dehumanizing effects of both "alienation in affluence" and "alienation in misery."[19] His perception stands valid even in the face of the conventional retort that "it's better to be rich and miserable than poor and miserable." What psychologists have learned of relative deprivation suggests that even if poor individuals harbor no hope for personal improvement, the mere vision of a "better" material life on the horizon gives them grounds for escape in fantasy. This form of relief is no longer available, however, to those whose material fantasies have nowhere else to go. Nevertheless, vicarious identification with "successful" people, especially those who have climbed the ladder from one's *own* social origins, provides an alternative outlet to fantasy.[20]

Hence technology is critical to development for four reasons: (a) technology is a resource and the creator of new resources; (b) it is a powerful instrument of social control even as it offers deliverance from underdevelopment; (c) it bears on the quality of decision-making to achieve social change; and (d) it constitutes a central arena wherein new meanings must be created to counter alienation, the antithesis of meaningful living.

It is no exaggeration to portray technology as a special universe of its own. Part One now explores this universe, especially those of its characteristics which bear directly on development.

Part One: The TECHNOLOGICAL UNIVERSE

Introduction

All societies express their practical skills in varied technologies, ranging from tools to building plans to preferred ways of diagnosing problems. Before development became a universal goal, most societies subordinated technological virtuosity to other values such as harmony with nature, ritualistic correctness, or the protection of existing authority structures. *Truly, the dominant and pervasive role played by technology is unique to modern societies.* Because it is linked directly to science, which has high prestige value, and thanks to its demonstrated practical utility, modern technology enjoys lofty status in the constellation of contemporary social values. And modern societies are the breeding ground of those technologies whose impact on development is greatest.

The Matrix of Technology: "Modern" Societies

Historians of science agree that special circumstances in Western Europe gave birth to modern science and technology.[1] The adoption by the West of the technological mentality as the primary mode of problem-solving is thus an untypical response given by one set of societies to common challenges issued to all by the forces of nature. To overlook this fact is to disqualify oneself from correctly assessing the impact of "Western" technology on other lands. But even without assuming that the Western response to challenges posed by nature is aberrant, one must acknowledge the important fact that the organization of society along dominantly technological lines is an event of comparative infrequency observed only in recent history.

All human societies display an "existence rationality."[2] Whatever may be their information-processing capacities and effective access to resources, all human groups devise concrete strategies which enable them to survive, to protect their identity and dignity, and to assert whatever freedoms they can muster over nature, over enemies, and over destructive social forces at work within their own boundaries. These strategies, taken as wholes, constitute their "existence rationality." Even communities which attempt to solve problems by consulting ancestors or propitiating fertility gods employ a strategy which is, in a true sense, *rational*: it rests on some proportion between available information and resources and perceived vital needs. The global diffusion of modern technology tends to standardize the "existence rationality" of all societies around specifically Western notions of efficiency, rationality, and problem-solving. This growing standardization is crucial because modernity is not the mere presence of factories but rather a new outlook on factories, whether or not they existed before. Thus in societies which are modernizing, not only factories but also the technological approach to life itself gradually come to be viewed as normal. Yet both factories and technology remain alien to experiential landscapes in nonmodern settings. Most value conflicts in technology transfers from "developed" to "less-developed" countries are traceable to two facts: (a) that modern societies are the historical matrix of the technological mind, and (b) that the very modernity of such societies makes of them breeding grounds for technology. The road to modernity necessarily passes through technology. Indeed, as John Montgomery writes, "Technology almost certainly offers the best hope of improving the quality of life in the developing countries."[3]

There exists, it must be added, a value universe which is proper to modern technology. The value content of this universe must be laid bare before one can properly assess the impact of technology transfers on "recipient" societies. The reason is that the value impingement of technology on these societies is twofold: some impacts are directly traceable to technology's inherent values and others to the channels through which technology circulates. The purpose of Chapter One is to analyze the ambiguous character of technology as a value universe.

1/The Two-Edged Sword

Technology is experienced as ambiguous for two reasons: (a) modern technology is simultaneously the bearer and the destroyer of precious human values; and (b) although it brings new freedom from old constraints imposed by nature, tradition, or ancient social patterns, technology also introduces new determinisms into the life of its adepts. This twofold character of technology needs to be "peeled away" in phenomenological fashion: beneath its surface characteristics, technology's ambivalent core must be unveiled.[1]

Technology: Bearer and Destroyer of Values

Four basic values are embedded in contemporary Western technology.[2]

The first is a particular approach to *rationality*. For the Western technological mind, "to be rational" signifies viewing every experience as a problem which can be broken down into parts, reassembled, manipulated in practical ways, and measured in its effects. The West is indifferent to what older traditions term *truth*. Speaking epistemologically, verifiability has supplanted truth.

I once attended a colloquium in Montreal at which a scholar from India made a striking statement which illustrates divergent views of rationality. One seminar participant observed that "the East needs the West and the West needs the East, but the East is not very dynamic, and its wisdom is static. What the West gave the East was the ability to change things and extract more out of nature, to mobilize human effort so that it can lead to greater results." The Indian scholar replied: "You Westerners insist on the need to live and think on the historical level. But those of us who still attach importance to the Hindu approach to reality have tended to live rather on the mythical level. We believe that the mythical level is no less real than the historical level." This person was in effect saying: It is ethnocentric for Westerners to assume that historical rationality is more real than other realms of cognition simply because its elements are amenable to direct observation. The Indian scholar further asked: "Does not the West too have a profound need for meaning, myths, and symbols?

Perhaps Western societies are so susceptible to ideological and symbolical manipulation because they operate in a vacuum of myths." An important value judgment emerges here: that Western technology is reductionistic in its approach to rationality.

A second value embedded in Western technology is its particular viewpoint on efficiency. Efficiency is a general relationship, and its dynamics can be laid bare by analyzing a specific expression drawn from industry, that is *productivity*. Production looks to the amount of final output; productivity, to some proportion between what is put in and what comes out. Like efficiency, productivity is gauged by comparing the product obtained with amounts of labor, capital, machinery, or time invested. Measures produce various comparisons between input and output, including composite evaluations of total input and total output. Whoever thinks about efficiency makes judgments regarding what to include and what to exclude from comparisons made. As economists would say, certain elements are treated as "externalities" to the efficiency calculus, while others are labeled "internalities."

To illustrate: Factory managers in the United States did not, until recently, include antipollution expenses in their benefit/cost calculations; pollution was treated as a mere "externality." The exclusion of this and similar values is easy to explain: Given the socioeconomic system within which Western technology matured, the production of goods by firms was treated in accord with a profit-maximizing calculus. Quite logically, therefore, important social values were systematically excluded. Behind this form of reasoning lies a mechanistic engineering mentality.

This mentality contrasts sharply with approaches to judging efficiency not yet dominated by Western technology.[3] One different mode of thinking is reflected, for example, in the behavior, as late as 1958, of two Bedouin tribes in the Sahara, the Ouled Sidi-Aissa and the Ouled Sidi-Cheikh, who spontaneously incorporated their Muslim beliefs and ethical values into their conception of efficiency.[4] Although they did not use the term, the "efficient" way to work, for them, was one that allowed them time to recite Koranic prayers seven times daily and to reduce their expenditure of physical energy during Ramadan fasting periods. Similarly, the "efficient" path along which to lead their flocks to pasture was not the shortest route but one that provided occasions for them to practice Koranic hospitality toward the poor along the way. These tribes had thus "internalized" religious and ethical values in their assessment of "efficiency." In fact, most non-Western societies continue to make a calculus of efficiency which internalizes religious, kinship, aesthetic, and recreational values in the performance of "economic" activities like agricultural work and hunting or fishing expeditions. "Modern" societies, in contrast, treat such values as "externalities." One cannot estimate the impact of

Western technologies on non-Western societies without adverting explicitly to this difference.

A third value of Western technology is its predilection for the problem-solving stance in the face of nature and human events. By definition technology is interested in getting things done; consequently, it breeds impatience with contemplation or harmony with nature. It also breeds impatience with the stance of indifference, passivity, or resignation in the presence of perceived problems.

At times developmental change may entail quite revolutionary political struggles. But the problem-solving stance favored by technology differs totally from the revolutionary "problematizing" stance. This important difference is repeatedly invoked in the writings of the Brazilian educator Paulo Freire.[5] According to Freire, one can know truly only to the extent that one "problematizes" the natural, cultural, and historical reality in which one is immersed. And how does "problematizing" differ from technocratic "problem-solving"? In problem-solving, an expert steps back some distance from reality, breaks it into and analyzes its component parts, devises means for solving difficulties in the most efficient way, and then dictates a strategy or policy. This approach, Freire contends, distorts the organic totality of human experience by reducing it solely to those dimensions which can be treated as mere difficulties to be removed.

To problematize, on the contrary, is to engage an entire populace in the task of codifying its total reality into symbols capable of generating critical consciousness and empowering them to alter their relations with nature and social forces. Problem-solvers who break reality down into parts remain outside viewers of that reality and are unable to grasp the totality surrounding them. But problematizers see themselves as part of that totality; in addition, that totality is itself subject to the influence of their own actions once they gain a new critical understanding of it. The reflective group exercise of problematizing a common social or historical condition is rescued from narcissism only if it thrusts all participants into dialogue with others whose historical vocation is like theirs, that is, to become transforming agents of their social reality. Thus do people become, in Freire's terms, "subjects" instead of "objects" of their own history. By adopting the "problematizing" stance, one perceives the totality of outside relationships. This is the prelude to viewing oneself as a social actor capable of transforming oppressive reality.

The problem-solving stance so essential to the technological mentality contrasts sharply, therefore, with two antithetical postures: the indifference to problem-solving of those who are passively fatalistic and the politicized "problematizing" of revolutionary change agents. This opposition explains why both "traditionalists" and "revolutionaries" find themselves ill at ease in the face of technology. Importing technology to "less-developed" societies can be so traumatic because it provides new legitimacy and new rewards to fragmenting modes of problem-solving.

A fourth value carried by technology is an exaggerated Promethean view of the universe. Natural forces as well as human institutions are viewed by adepts of technology as objects to be used and manipulated; indeed, the value of their existence is equated with their usefulness, that usefulness in turn being rendered or conferred by a force which enables men to control and change nature. technology. In contrast, most traditional societies presuppose some kind of harmonious compact with nature and its forces and seek to minimize the damage done to life. Plant life, for example, was so precious to American Indians that when a man stepped on a plant or bruised a stalk of wheat it was his duty to make a ritualistic invocation to nature to express regret for violating life. He felt a deep kinship with the nature he was hurting and a sense of responsibility for the harm wrought, even if his actions were necessary for his survival.[6] This stance is the polar antithesis of the exploitative Prometheanism which so deeply characterizes Western technology.

Technological innovators do not intend, of course, to destroy pre-existing values; their overt aim is merely to solve some problem more efficiently, to produce goods or provide services according to different standards of quantity or quality than before. Nevertheless, simply by acting as innovators, they cannot avoid tampering with prior values. Worse still, they shatter the fragile web which binds all the values of premodern communities into a meaningful whole.

In localities only slightly affected by Western technology a close nexus still exists between normative and signifying values, between rules for action and symbols conferring meaning. "What ought to be done" in such domains as family relations, work, commercial exchange, and dealings with leaders is intimately related to the images society adopts to explain to itself the meaning of life and death. This nexus is absent, on the contrary, from developed societies, where no unifying vision of life's total meaning is shared by all and where, in fact, the opposite is true: society exhibits great tolerance toward a plurality of significative values. To illustrate: A devout Mormon businessman in the United States obeys the same professional ethic as his agnostic counterpart. Yet, although the norms guiding their respective professional behaviors are the same, the ultimate significance attached by each to these norms differs. The Mormon works hard and remains honest in order to "do God's bidding," whereas the other's motive is simply to "play the business game fairly" and avoid losing his good name. The gap separating behavioral norms from deeper symbolic meanings poses few overt problems in industrialized societies. The opposite is the case, however, in transitional societies. A fuller understanding of that gap is important, therefore, if we are to gauge the full impact of imported technology on their social values.

Whereas in less-developed societies a high level of integration exists between normative and significative values, economic activities are fragmented. A large number of production units—individuals, families, or villages—operate quite independently of others, and little coordination of effort or specialization of tasks is required. The opposite condition prevails in "developed" areas. There the basic symbols which explain history, life, and personal destinies have no link to norms for action, but economic activity is so highly integrated that the autarchic subsistence of small units becomes practically impossible. Ultimately, the importance of the nexus between norms and meanings lies in this: In traditional societies work is a cosmic act; in developed societies, a specialized function.

Through modern technology traditional societies receive change stimuli which directly challenge their *normative values.* These challenges take the form of models for doing things differently—planting crops, educating children, or practicing hygiene. Likewise, new goals of human effort are suggested: to earn cash, to build a "better" house or eat more food, to gain greater mobility. Proposals for change raise an existential question: Should members of society continue to act as before or modify their behavior norms? Because traditional behavior norms are founded upon, and derived from, a given universe of explanations, however, to challenge norms is *ipso facto* to attack the underlying belief system. I recall witnessing the rapid deterioration of a father's authority over his sons during the Algerian war of independence because food scarcity forced his sons to take salaried jobs with a French road-construction gang. And his degree of control over his wives lessened because the French government organized in 1958 a referendum to induce the Arab populace to keep the French in power. The French pressured Muslim women into voting in this referendum. Even in small Saharan towns, the referendum campaign had a shattering impact on the Islamic values of the community. Many women had never before taken part in any activity outside home and family. In the very process of modifying behavioral norms, the referendum attacked the symbolic system of the society: it shattered the nexus between normative and significative values.

Once this nexus is shattered, affected societies must choose between two demoralizing options. The first is to adhere to ancient signifying values even if these are contradicted by behavior norms which increasingly determine daily practical activity. Such fragmentation is psychologically damaging to people accustomed to attaching cosmic meaning even to simple actions carried out in the home, in the field, or on the pathways. Because serious identity problems are posed by inducements to new behavior, change frequently tends to be either rejected outright or embraced too uncritically. Both reactions produce damaging consequences. Although certain Western writers fondly praise "achievement orientation" and the "spirit of initiative,"[7] in many societies both are moral aberrations as reprehensible as theft and criminal neglect are in others. In short, the first option open to "transitional" societies is to live in a state of value fragmentation in which their cherished meanings are daily violated by new rules of action.

The second choice theoretically open to them is to fashion a new coherent nexus between meaning values and rules for action. Such a synthesis is practically impossible, however, in the short term. How can groups experiencing modern technology for the first time quickly create a new synthesis of meaning and practical norms when advanced countries themselves, after two centuries of familiarity with technique, have proved incapable of devising a wisdom to match their sciences? Western societies do not provide a valid framework of broad goals within which to evaluate science or technique. As Danilo Dolci writes, "We have become experts when it comes to machinery, but we are still novices in dealing with organisms."[8]

Societies initiating themselves to modern technology lack the long familiarity with science and technology which might enable them to make a new synthesis between these and their ancient wisdoms. And they have no realistic hope of preserving unity in their world of values by uncritically assimilating new techniques. Therefore, they are condemned to social disruption unless they can successfully involve their entire populace in decisions regarding tolerable value sacrifices to be made in accepting proposed change.[9] No less necessary is a different mode of impingement on affected communities by external change agents, who need to learn how to respect the inner core of a group's "existence rationality." Values which are essential to a group's identity and dignity are the core of its existence rationality or strategy for survival. At stake in social change are values more basic than divergent perceptions of rationality, efficiency, or problem-solving. Vying for supremacy are two competing visions of the "dynamism of desire" which elicits personal energies required to "make society work." Desire is the arena where acquisitive urges of individuals are regulated by norms of social legitimacy. One best understands the dynamism of desire by comparing limits placed upon desires in pretechnological and technological societies. More specifically, it is necessary to examine the changes which occur when technology, bearer of a new dynamism of desire, impinges on societies.

Technological levels prevailing in non-Western societies did not allow them to achieve high degrees of productivity, that is, to extract a high ratio of new wealth to inputs of effort or invention. As a result, these societies aggregated only limited resources for consumption by their members. Both symbolic and normative value structures had to accept these constraints as givens. Resources were neither abundant nor inexhaustible, and little likelihood existed that they could increase significantly within the lifetime of one generation. Accordingly, social norms governing access to, and use of, resources had to be based on one of three values: equity, hierarchy, or priority needs. All three dictated a curbing of desire and of the acquisitive spirit. Were the brakes on desire removed, individuals would make dangerous claims on a static and limited pool of resources. To legitimate personal acquisitiveness could ruin a hierarchic social system or shatter the solidarity binding kin, one to another, in patterns of reciprocal obligations. To foster the acquisitive spirit of competitive individuals by legitimating it would produce what game theorists call a zero-sum game, in which any material gains obtained by competitive individuals would be won at the expense of those remaining in dire need.

This image of a normatively limited dynamism of desire depicts, with some qualifications, the conditions found in most non-Western technological societies. Upon its arrival technology becomes the vector of the virus of acquisitiveness, thereby shattering the delicate balance between social restraints on desire and effectively available resources. Technology undermines the norms of need-satisfaction which have allowed countless societies to survive. Yet it is far easier to generate new aspirations than to augment resources. But once normative constraints on desire are removed, even if resources do increase thanks to technology, surviving social institutions and normative structures can no longer assure that increases are not appropriated by a few at the expense of the many.

This traumatic disturbance of the finely calibrated dynamism of desire in non-Western communities occurs independently of the intentions of those who channel technology. Technology attacks the principle of cohesion which wove the value universe of pre-industrial societies into a unified fabric. It also undermines the view of nature such societies have, the meaning they assign to work, to time, to authority, and to the very purpose of life. Under the assault of technology, work can no longer be seen as sharing in the creativity of nature or as expressing cosmic relationships; it becomes the mere performance of a task whose only meaning comes from the external rewards, usually monetary, attached to it. Similarly, time can no longer be lived as a rhythm of maturation cycles but must instead be endured as a succession of atomistic moments to be used efficiently and profitably. Kinship and other intimate relationships become subordinated to criteria of performance or power. Authority itself, formerly legitimated in stable patterns by meaningful symbols bringing vicarious satisfactions even to those lacking power, becomes negotiable and subject to laws of "competition."

The process which produces these changes is one of pervasive "commercialization"—of friendship, of procreation, of love, of partnership. That deep social bond which writers like Mauss, Dillon, Perroux, and Titmuss[10] call the "gift relationship" is irrevocably destroyed: all exchanges are henceforth governed by the law of interests. Creative acts dictated by love, spontaneity, or esthetic fervor—what existentialist philosophers call "gratuity"—and the fulfillment of "social obligation" in the twin senses of duty and the creation of personal "bonds" are banished from social intercourse. In a word, modern technology is a powerful vector of the acquisitive spirit.

The major impact of technology, however, is that in the final instance it reifies all values. Technology bears within itself powerful determinisms. There is no need to review here debates on the "technological imperative" occasioned by the publication, in English, of Ellul's *The Technological Society.*[11] Yet no inquiry into the role of technology in development can ignore this paradox, that while technology frees societies from old constraints it creates new determinisms as well.

Freedom from Old Constraints, or New Determinisms?

An inner force drives technology to render actual everything which is possible: this is the redoubtable "technological imperative." Lord Ritchie-Calder comments on a contract once given to the University of Michigan to study the feasibility of a planned city of 60 million to be built north of Bombay:

> You know what happens when enthusiasts get busy on a "feasibility study"—They prove it is feasible! And once they get to their drawing boards, they have a whale of a time. They design skyscrapers above ground and subterranean tenements below ground. They work out how much air can be spared and hence how many cubic feet of breathing space is required for a family. They design "living-units" which are just hutches for battery-fed people. They design modules and clamp them together, pile them up like kindergarten bricks. They lay on water and regulate the sewage—water and sewage, in this case, for 60 million people—on the now well-established principles of factory-farming. And when they have finished they can prove that this is the most economical way, cost-efficiencywise, of housing people. I hope that project has been scotched; I did my best to convince my friends, with influence, in India. I asked them, for instance, how many mental hospitals they were providing for the millions who would go mad under these conditions.[12]

The problem is general: means tend to usurp the place of ends: processes express their own dynamisms apart from goals they are meant to serve. Proving that any idea can be translated into an artifact constitutes for many engineers or chemists a challenge they find irresistible. Momentum builds up within research institutions for them to do something mainly to prove that it is possible. And much irresponsible technological tinkering is encouraged by the benevolent attitude towards change which prevails in "developed" societies, whose general bias favors the view that what is new is necessarily better. Technological researchers embrace this bias—and are rewarded by society for their efforts! This bias gives them a vested interest in perpetuating the "technological imperative," that is, the tendency of technology to impose itself independently of larger purposes. A few moderns may, it is true, remember that Leonardo da Vinci refused to communicate the plans of certain inventions (e.g., submarines) to princes or to develop working prototypes after his designs had been perfected.[13] They may also praise the wisdom of ancient Chinese rulers who chose to use gunpowder *not* for warfare but merely for fireworks on ritual days. In past centuries, however, technology had not yet become the object of a near-idolatrous social cult. Thus technology's "imperative" is traceable not to anything intrinsic in technology but to the infatuation of contemporary human beings with their own creativity.

Ellul is often accused of overstating the autonomy of technology. His retort is that technology *need not be* deterministic, simply that in fact it operates powerfully in the direction of determinism.[14] Unless those who would harness technology to humane tasks correctly assess the strength of the technological imperative, they will be unable to gain mastery over change processes. The history of military research in the United States attests to the mystifications surrounding conventional arguments for technological expenditure.[15] Champions of expanding military research often claim publicly that the main benefit derived from their efforts is the spinoff application of their technological findings in civilian sectors. Such spinoff is, however, modest in scale compared to total investments. Even in purely economic terms it would make more sense to focus research and development energies directly on civilian problems the solution to which is anticipated rather than to hope for spinoff from military research. Yet potent vested interests promote such technological pursuits, thereby providing employment to countless technicians. Notwithstanding the claim that it brings a "competitive edge" in weaponry, military technology is an insatiable Moloch: "technological perfection" is never reached. On the contrary, competition dictates that "it is feasible; therefore, it will be done." Ultimately, the technological imperative can be moderated only if the technological cost/benefit equation is radically altered. One must internalize new externalities.

Externalities

According to economists, an externality is any value or consideration which does not enter a cost calculus.[16] Why did dramatic crises have to erupt before the US public began to understand that factories or weapons dangerously contaminate the atmosphere? Because survival and clean air were treated by policy-makers as "externalities"; that is, for the specific purpose of making production decisions, such values were deemed irrelevant. But the social, psychological, and ecological costs of any economic or technological activity are never irrelevant; they determine the very desirability of that activity. Numerous values formerly treated as externalities need to be internalized if sound social decisions are to be reached.

The principle of responsible internalization is illustrated in the case of auto safety. So long as saleability and luxury appeal were treated as major "internalities," auto designers could treat safety as a mere "externality." They could do likewise with fuel economy if they could plausibly assume that gasoline would remain plentiful and cheap. Once fuel economy became paramount, however, and public pressure grew to provide greater safety in vehicles, new constraints became "internalized," leading not only to different designs but also to a new economic equation measuring costs and benefits. The lesson is clearly that *the technological imperative will lead to excessive determinism unless resistance to determinism becomes an internality in any decision about technology.*

Once the notion of countering determinism becomes an explicitly internalized goal, planners will conclude that certain technological applications must not be adopted and that others should be slowed down or redirected. Technological research and development will continue, to be sure, but will not be allowed to proceed unchecked on the assumption that they insure nothing but unequivocal benefits. Most decision-makers lack the wisdom to match their sciences. Therefore, the beginning of wisdom consists in not rushing headlong into further technological pursuits regardless of social and human costs. At stake, finally, is the capacity diverse societies possess to absorb technologies which are simultaneously creators and destroyers of social values.

Resistance to determinism is not the only externality needing to be internalized. Other major developmental values are also internalities: equity, cultural diversity, ecological health, and reduced dependency. Societies can begin to harness technology to proper ends only if they understand that technology is simultaneously a universe to be created and an artificial context for their economic and organizational rearrangements. It is so difficult to control technology or to dominate nature without damaging it because the Promethean spirit is so powerfully seductive. The domination this spirit promises deceives people into treating technology as its own justification.

If moderns continue to treat their own creation, technology, as they have treated nature in the past, they will not escape technological determinism. Indeed, adopting a Promethean stance towards technology forces one to rely on still more technology in order to control technology itself: this is the "technological fix" mentality. Men have used technology to conquer nature. Had they respected nature in the past, however, they would have devised technologies quite different from those they actually produced. They will make similar mistakes in their efforts to moderate technological growth unless they repudiate the stance of untrammeled exploitation. Like nature, technology cannot be controlled with impunity unless it is first respected, for technology—like nature—imposes its own rhythms. Machines, tools, and computers should not be made to do more than they can do, lest they impose their logic on those who tend them and mechanize their makers. Analogies abound in the arts. Sculptors respect their tools—chisels and hammers—and musicians theirs; that the tools and instruments are of human manufacture is no excuse for abusing them. One can learn to respect technologies by designing them to last and to express esthetic as well as functional values. Such a respectful attitude is the antithesis of the cult of technological obsolescence and of pure functionality which presently dominates. Indeed "developed" societies have ravaged so much of nature's beauty that they cannot live without new forms of technological beauty to take its place. A minority of architects and designers, it is true, has always advocated making beauty an "internality" in the design of "functional" objects—dwellings, furniture, office equipment, and tools. In the main, however, their efforts have been viewed by manufacturers and by the public as luxuries. But simplicity, beauty, and durability in everyday technologies are *not* luxuries: they are no less important than utility or efficiency.

A liberating imperative must oppose determinism by making technological design the choice arena where social-value externalities get internalized. But if the effort is to succeed, efficiency itself will have to be redefined.

Redefining Efficiency

The notion of efficiency which governs decisions in hundreds of work sites is that born in the engineer's mind. Efficient operations are measured by comparing inputs in energy, time, or money with quantified outputs. In the words of one dictionary, efficiency is "the ratio of the useful energy delivered by a dynamic system to the energy supplied to it." Because technology has operated largely in closed circuits where engineering values were unquestioned, it has become the arena par excellence where efficient solutions are defined without regard to social externalities. The bankruptcy of this procedure, however, daily grows more apparent.

At least three values must now be internalized in any efficiency calculus: the abolition of mass misery, survival of the ecosystem, and defense of the entire human race against technological determinism.

Noteworthy examples of organized technological activity in which broad social externalities are internalized in decisions already exist. One is the national policy of the People's Republic of China, which overtly incorporates political criteria in its choices of technologies. "Politics is in command" is the watchword of the process. Politics is interpreted to mean the constellation of basic values to be fostered: social equality, the diffusion of revolutionary consciousness, the "serve the people" ethic, and respect for those who perform menial tasks. (More will be said in a later chapter on the Chinese value structure as it affects technology.) Other examples of redefining efficiency are the efforts made by advocates of "radical," "intermediate," and "soft" technology in many societies.[17] At issue here is more than a new definition of efficiency; the struggle is against the tyrannical hold which the engineering cast of mind continues to exercise on the thinking of designers and technicians. Conceptually, what is required is not unclear, but it is difficult to implement. Huge vested interests, not the least of which is the intellectual security produced by two centuries of thinking in familiar patterns, stand in the way. Most moderns simply do not know how to be efficient without destroying the environment, alienating workers, or reinforcing technological determinism. Out of habit they judge the efficiency of machines and processes by systematically excluding important social values. The balance is difficult to redress because a host of problems press for solution on the old terms. This is why efficiency needs to be redefined via political consultations which bear directly on value priorities and the allocation of social costs. It is no longer correct to label some procedure efficient if it exacts intolerable social costs, proves grossly wasteful of resources, or imposes its mechanistic rhythms on its operator.

Comparisons must be made between *total inputs* and *total outputs* in the functioning of any technology, for technology itself, as presently utilized, is a large part of the very problem to be solved. It is obviously futile to look to some new technological priesthood of the wise for salvation, as Saint-Simon did a century ago. What is needed is a new breed of technicians and engineers who, if they are not themselves philosophers, are willing to trust the philosophical judgment of common citizens in the political arena. Paradoxically, decisions about efficiency will need to be conducted in a consultative mode which, at first glance, seems inefficient. But a new balance must be struck between obeying the "inner" efficiency demands of technology's logic and the "external" demands imposed by the higher logic of social-value enhancement. The difficulty to be faced is like that which confronts agricultural extension agents and other pedagogues of social change.[18] If their objective is solely "to get the job done," they will tire of waiting for peasants or "underdeveloped" counterparts to be able to "do it right." Yet if extension agents merely do the job themselves and go away, they doom their own efforts to failure by leaving no one behind who knows how to "do it right." Or, in situations of more formal education, where pupils have been taught to perform tasks by imitation or rote, the "getting the job done" approach falls prey to mystification on two counts. First, learners will not assimilate the reasons why things are done in the technological mode, and second, they will not be able to integrate technique into their own universe of values. The lesson is simple: time seemingly "wasted" in consultations on how to "internalize" value "externalities" is, in fact, no obstacle to efficiency. The opposite is true: time spent thus is necessary if utilizers of technology are to become masters of their tools. But, one may plausibly object, even if this objective is sound, insuperable problems of measurement are involved. After all, are we not totally ignorant of how to compare *human* costs with reified problem-solving outputs?

No answer can be given except by tracing the connection between methodological and value questions. Social scientists studying social change often invoke difficulties of measurement to justify their posture of ignoring the value questions. An analogous problem is encountered in the realm of social auditing.[19] A parallel difficulty is faced in redefining efficiency in practical ways. Measures for comparing qualitative value costs and quantitative outputs are doubtless lacking, but no less debilitating is the influence of the attitudes held by engineers and their consuming public regarding changes in the efficiency equation. This obstacle blocks change in many realms: relations between medical doctors and patients, between educators and students, and between experts of all types and the supposed beneficiaries of their ministrations. The redefinition of efficiency would strike a damaging blow to many vested interests. It would first undermine the claim that "experts" know more about important issues than nonexperts. It would, moreover, require of producers that production be rendered not merely socially responsible but positively enhancing of larger societal values. These demands may reduce the benefits of producers or limit the maneuvering room they enjoy to achieve an edge over competitors.

The past history of technology renders the task doubly difficult. Had modern technology evolved from the laborious trial and error of peasant masses and been subsequently refined by technical elites, a socially responsible efficiency calculus might be founded on appeals to justice. But the history of technological development renders any such appeal implausible, for technology was historically devised by the creative efforts of a small number of elite mavericks. Now, however, in the name of redefining efficiency, their successors—the class of technicians—are asked to dilute their influence early in the decision-making process. Most technical "experts" will not willingly share their hard-won, and recently acquired, political power with the general public. To suggest that they do so is to call for a new kind of Copernican revolution![20]

Disagreements exist as to the best institutional *locus* of efficiency. Many who are alarmed by waste in governmental agencies equate efficiency with the private sector. Others, reflecting on the private sector's uncanny ability to rationalize private goals under the banner of public service, look with suspicion on any private definition of efficiency. The answer lies not in absolute dichotomies, however. Enterprises in both sectors must create new modes of operating efficiently, simultaneously solving problems in the conventional style and optimizing social values hitherto externalized but now needing to be internalized. The role played by competition in legitimating growth is crucial, for efficiency can never be achieved without some form of competitive emulation and accountability. One recalls how important have been Stakhanovite emulation in the Soviet Union and its counterpart in China, the public posting of production performances of work brigades singled out for praise. There are, of course, certain rock-bottom economic constraints; good social intentions are no substitute for the practical arts of efficiency. But firm managers and designers of technology will need to explore ways of becoming integrally efficient—that is, of producing efficiently while optimizing social and human values. This they must do with as much passion, singlemindedness, and practical sense as they now devote to making profit or creating new products.

The central value profile of the technological universe has now been drawn, revealing technology as a two-edged sword which is at once bearer and destroyer of values. But its dual character extends to another domain—that of freedom and necessity. The same technology which frees its utilizer from constraints imposed by nature, traditional norms, or cultural taboos also introduces new patterns of necessity. And because technology tends to blur dividing lines between means and ends, it rapidly builds up vested interests in testing all its latent possibilities, with sublime indifference to costs in sacrificed cultural values or personal suffering.

Yet one cannot grasp technology by looking only at its static qualities; technology is a dynamic force, constantly evolving and interacting with broader forces of social change. The following chapter, accordingly, inquires into the dynamics of technology.

2/The Dynamics of Technology

Technology, far from being an inert deposit of practical knowledge, is itself a rapidly evolving system operating within larger systems also undergoing dynamic transformation. The radical instability which characterizes technology's matrix, modern societies, perfectly embodies the imagery made famous by Heraclitus—"all things flow." The very texture of life in such societies is swift, perpetual, and ineluctable change. Pervasive change creates expectations of further change and conditions people to view innovation as a value for its own sake, quite apart from considerations of intrinsic merit. No more congenial setting could be found for the development and continued growth of technology.

Technology has become, for moderns, the functional equivalent of nature for primitives. The present chapter briefly explores how technology is a kind of "second nature" and identifies various sources of technological dynamism. These include the competitive structures operative in the developed world, capitalist and socialist; the interaction among basic value choices in any society, its preferred development strategies, and its approach to technology; and the "sequence of dependency" which marks relations between rich and poor countries in arenas of international exchange.

Technology as Second Nature

In 1954 Ellul wrote that "no social,...human, or spiritual fact is so important as the fact of technique in the modern world."[1] Ellul's term *technique* is roughly synonymous with *technology* as used in the present work. Moreover, one central assertion made herein parallels a major theme of Ellul's, namely, *that technology has replaced nature as the context of societal perceptions and decisions.* For modern societies as for transnational societies now facing its challenge, technology, not nature, is the boundary against which possibilities must be measured.

In earlier ages humans experienced and interpreted reality against the backdrop provided by nature; indeed, a dominant part of the reality they perceived was nature itself. Their plans for survival and physical activity depended on the regularities or caprices of natural forces—heat, cold, wind, rain, seasonal cycles. That artificial sunlight called electricity did not yet exist; neither did the man-made bird called airplane, the artificial eye known as camera, or the surrogate mountains and forests we call skyscrapers and cities. Both in perceptual time and in importance nature was primary. The events which most dramatically influenced individual lives were natural: rainfall, droughts, floods, storms, and good weather were woven into the tissue of births, puberty rites, and deaths. Compared to natural forces, whatever an individual, family, tribe, or village might affect seemed puny. One planted and weeded, of course, but crop success depended above all on the weather. Society mobilized young males for hunting, but a sudden storm could chase the game out of reach. One built houses, but torrential rains could bring them down in an instant. Rivers were benevolent or destructive, winds capricious, the seasons themselves uncertain. Economic outcomes, no less than social harmony, depended largely on how nature behaved. Inevitably, all societies felt bound to render prodigal ritualistic homage to nature's supremacy through symbols, festivals, and personal obligations. More significantly, norms prescribing social behavior were designed to respect limits set by nature.

In modern societies technology has displaced nature from center stage. This is also the case in premodern societies increasingly caught up in processes of change. The impersonal forces to which society must now relate are those created by technology: electricity and other forms of artificial energy; machines able not only to perform myriad tasks but to design them as well; and decisional techniques which interpret and manipulate human life at every level. Such technological events as building roads, operating large factories, administering new cities, and disseminating radios or contraceptives *en masse* leave people feeling far more powerless than do such natural happenings as lightning, storms, or floods. The only "natural" world which a growing number of the world's inhabitants directly experience comprises the artificial mountains, streams, and forests built by technology: skyscrapers, faucets, pipelines, and cities.

Technology has become, for many people, the significant reference point against which possibilities and constraints must be measured. Therefore, moderns pay their ritualistic homage to technology instead of to nature. Within "developed" cultures it has now become mandatory to praise technology and to endow it with esthetic legitimacy (formerly reserved to nature and to gods) by glorifying it in art, music, and worship. Beyond ritualistic tribute, however, moderns must guide their actions by what technology can do, should do, and perhaps even impersonally wants them to do.

To probe the full implications of technology as second nature, particularly its penetration deep into modern psyches, would require writing an ambitious, albeit an exciting, book. My present task lies elsewhere, however: the metaphor of "technology as second nature" is invoked here solely to set the stage for a discussion of those systemic properties of technology which are germane to development. Of central importance is the ambiguity inherent in technology both as social reality and as artificial nature. To what extent does technology determine development images, strategies, and accomplishments? In one important respect technology is unlike nature, for it changes very rapidly. Its mutations are recorded in years and decades, not centuries or millennia. Is it, ultimately, the potent dynamism inherent in technology which explains its dual impact on human values, simultaneously destroying and creating them? But does technological progress necessarily presuppose speedy change within the larger society? One cannot answer this question without first understanding how technology propels, and is in turn propelled by, the engines of economic competition. It is no exaggeration to regard technology itself as the key to the "competitive edge."

The Competitive Edge

In 1705, a full seventy-one years before the appearance of Adam Smith's treatise *The Wealth of Nations*, Bernard Mandeville wrote *The Fable of the Bees.*[2] Since then the notion of progress in the West has been associated with quantitative, and particularly with economic, growth in a framework of socially sanctioned competition. More recently the cult of growth has extended to technology. Technology is presumed to progress or improve if it grows—in size, influence, areas of penetration, and number of new products it creates.

During interviews I have often asked corporate managers and research directors: "Why are research and development (R&D) so important to you? Do its benefits justify such huge expenditures?" Almost uniformly their answer has been: "We have to keep up because technology is always changing. And it changes constantly because it gives those who possess it a competitive edge which confers a decisive advantage over others in arenas of economic competition." A similar assumption is made by many government officials in Third World countries: namely, that technology must keep growing if it is to serve the cause of development. This twin legitimation—on grounds that corporations need technology in order to remain competitive and that poor countries need it in order to develop—decisively affects thousands of corporate investment decisions and governmental choices regarding R&D. These decisions, taken cumulatively as a systemic whole, transform technology into a compulsive growth industry. Champions of technological expansion rarely pause to ask whether quantitative growth is better than steady state (qualitatively distinct from stagnation) or whether their chosen pace of acceleration does not render the affected social systems unmanageable.

Within industrialized countries social critics now bemoan the absence, in expert policy-makers, of wisdom to match their science. Perhaps one reason why wisdom is so scarce is that technological applications of science have been made to grow compulsively in order to serve the cause of competition regardless of social costs, tangible and intangible. The notion of "competitive edge" merits further analysis because it goes to the heart of the evolutionary dynamism of technology itself.

Certain development writings imply that science and technology are the common patrimony of mankind and that the Third World enjoys advantages in being a latecomer on the scene of technological modernization. The Third World, we are told, can take technological shortcuts. Yet technology is not a free, but an economic, good sold dearly to those who can pay for it, not to those who need it most. As Lord Ritchie-Calder explains:

> It is true that one does not have to re-invent the wheel in order to ride a bicycle. It is true that each country that undertakes the modernization of its economy relies partly on the heritage of others. It is also true that there is a great deal of knowledge and know-how freely available for transmission from one country to another but many of the less developed countries do not know how to go shopping in the supermarket of science (Nobel laureate Patrick Blackett's phrase) nor how to get the free samples of generally available technology. The term "transfer" in this sense is a euphemism because technology and know-how is being bought and sold like a commodity, but there is no world market nor a world exchange nor world prices for technology. The "latecomers" in this case are like spectators arriving at the last moment at a cup final and having to buy tickets from speculators at excessive prices.[3]

Technology may be the most vital of economic goods because it can generate new wealth faster than other productive assets—capital, labor, natural resources, or favorable location. If new wealth is the golden egg, technology is the hen that lays it. The institutional capacity to generate technology permanently and in self-sustaining fashion constitutes a priceless asset. A research and development laboratory is but a special kind of factory which produces an important capital good known as technology. Neither the factory itself, not its output, technology, is a free good or the common patrimony of mankind.

Unless exchanges are subsidized, technology must be paid for by the buyer. The proper arena for its circulation is some local, regional, national, transnational, or global market. Although much of it is proprietary knowledge, technology tends to circulate faster and easier than most other capital goods, indeed, than many consumer goods themselves. This greater mobility is explained by the relatively intangible nature even of technologies which are incorporated in a "package" of goods or services. What is worth noting here is that technology circulates, if at all, within arenas of economic competition in the production and provision of goods, products, and services. Thus is technology caught up in the dynamics of competition. This fact leads directly to a question: To which stimulus does competition itself respond in modern economies?

Competition is fueled by incentive structures which reward those who are the first on the list at meeting effective purchasing power and its equivalents. Goods are produced and supplied by various enterprises—private, public, or mixed. Their supply role is meaningless, however, unless matched by a vigorously exercised parallel demand function. Whether producers are decisively stimulated by the lure of moneys held by purchasers or by the rewards that come from those who wield effective power to set targets, competition remains the basic ground rule of economic activity. Within capitalism, competition as response to effective buying power enjoys priority as the motor force of mobilization for production. Under socialism, on the other hand, competition—or "emulation," as it is more generally termed—responds to motivations based on political, ideological, and bureaucratic interests. Even state-owned enterprises must compete among themselves to be awarded contracts, to gain access to sources of material inputs indispensable to production, and to meet targeted quotas in time to avoid punitive measures. Under both systems, it is competition in the arena of production which dictates the behavior of individual production units, even though these units respond to diverse stimuli which play the role of inducing and rewarding production in a competitive mode. Thus, although considerable differences separate the two systems, both place the quest for a competitive edge at the center of enterprise planning. Yet why, one may legitimately ask, are enterprises so important?

Not only are enterprises the main producers and consumers of technology, but they also rely heavily on their technological abilities to gain or preserve any competitive edge they may enjoy. Nonetheless, significant differences among them are discernible in arenas of competition. Where enterprises, be they private corporations or state agencies, function as monopolies or oligopolies, they can indulge in the luxury, at least for a time, of being indifferent to those marketable "qualities" of their output best supplied by technology, packaging, or advertising. In theory, the very monopoly held by such firms would render them immune to the challenge of competitors did not practical constraints dictate otherwise. But after all, even state monopolies must meet production goals, quality standards, and minimum general-performance levels; if enterprise managers fail, the government planners—who are their masters—pass judgment on them on the basis of the performance of *comparable* enterprises in other countries or in other sectors of the same national economy. And as long as some competitive sector exists in which winning or keeping an edge is important, competitiveness rules the arena within which enterprises play out their roles as producers of goods. In the Soviet Union and other socialist national economies, a broader domain of competition prevails—that between the respective abilities of capitalist and socialist economies to "deliver the goods"—and competition creates pressures even upon monopoly enterprises in the socialist sector to gain a competitive edge founded largely on technology. Within capitalist economies, in contrast, monopoly positions are ephemeral and precarious by definition, and oligopoly advantages are even more so.

To summarize, in uncontrolled classical "free" markets, the competitive edge is essential to the survival and prosperity of enterprises. In controlled markets (monopoly and oligopoly situations), although the competitive edge is *relatively* less crucial on purely economic grounds, external considerations dictate some degree of competitiveness. What results is a universalized drive to "keep oneself competitive" by keeping abreast of technological innovations.

For purposes of this book it is worth recalling that most exchanges take place in market arenas, because even "nonmarket" transfers prove, upon examination, to be disguised market exchanges between a seller and a purchaser subsidized by some third partner. Because the competitive arena remains dominant, individual providers of goods feel obliged to seek some kind of "competitive edge." Consequently, even enterprises enjoying monopoly or oligopoly advantages constantly experiment with new technologies, new products, new packaging, and cheaper production processes. They seek two goals: to protect their position from encroachment by outsiders and to prepare themselves to enter other arenas where they do not (yet) enjoy control or dominance over the market. Indeed market conjunctures change quickly, and even monopolies are vulnerable to shifts in product life cycles and altered demand structures.[4] Other sources of change likewise affect control over markets: the pressure of governments and political militants on monopolists; shifts in buying power (either quantitative changes in monetary power or compositional shifts in consumer markets); and competition from enterprises eager to "break" the monopoly or share in the advantages of oligopoly. One lesson stands out: Complacency kills privilege. Accordingly, what business theorists term a "defensive" posture aimed at avoiding losses of privilege turns out, upon closer examination, to be no less direct a stimulus of competition than is an offensive stance aimed at gaining profit or privilege. Great latitude for aggressiveness is found whether firms pursue absolute profit gains or relative gains in their "share of the market."

One question—What provides the competitive edge?—has long

puzzled theorists of the corporation. Frederick Knickerbocker traces it to a firm's oligopoly position.[5] But whence comes the oligopoly position itself? Its ultimate source is some competitive advantage expressed as a new product, better packaging, cheaper production processes, higher or more standardized quality, the ability to use alternative materials, or favorable access to special market slices. All these advantages, except the latter, are traceable to technology, which enables one firm to achieve these relative gains over others. Technology also enables competitors to wipe out the "edge" others enjoy and themselves become "competitive."

Yet one must not suppose that all technologies are equally stable. A few examples may prove helpful. Technologies used by shipbuilders[6] or dredge constructors change more slowly and less drastically than those utilized by makers of precision instruments or computers. Similarly, technologies acceptable to firms extracting minerals evolve more slowly than those utilized by manufacturers of carbon black or processors of petrochemicals, or even by those who refine or otherwise process extracted ores.

To the important question, "Why can certain activities remain competitive through the utilization of relatively stable technologies whereas others require ever-changing technologies?" several partial answers suggest themselves.

(a) *Scale constraints* explain some differences. In any activity requiring huge sums of capital and large basic infrastructures, actual and potential competitors cannot enter the arena quickly. Even assuming that competitors possessed superior technologies, they would lack other requisites for quickly translating their superiority into actual competing enterprises. Not surprisingly, the large size of the investment made by the initial firm in the arena makes it itself cautious about altering its equipment and/or processes before amortization has been effected. Although size alone does not *impede* rapid technological dynamism, it *slows down* the rate of change.

(b) *The nature of the product* also affects the relative stability of the technology used. If, for example, relative to the very nature of the materials used, fabrication is unsafe, materials are awkward to handle or transport, expensive or difficult to package, then a powerful stimulus exists to induce actual and potential competitors to make technological changes, for the reason that technological breakthroughs on these fronts confer immediate marketable advantages. On the other hand, if materials or processes are safe and easily handled, lesser inducements exist to concentrate R&D in search of improvements. Even when such improvements are made in laboratories, they cannot quickly be translated into sizable market advantages. In contrast, in domains where concerns for assuring health and safety are paramount—pharmaceuticals, chemicals, medical products, volatile or inflammable materials—technologies are highly unstable.

(c) *Luxury goods* and their equivalents also are biased toward rapidly changing technologies.[7] Demand for these categories of goods depends heavily on subjective factors easily manipulated by advertising. Design, shape, color, model variations, and packaging take on great importance in determining the size and location of the market for such goods. A powerful incentive exists for enterprises to engage in R&D because, by definition, potential buyers are conditioned to desire frequent changes. By applying this criterion to diverse technologies, one understands why bread-baking technologies tend to be less varied and more stable than those used in making cookies and those used to make screwdrivers more stable than those for electric lawnmowers, power saws, or phonograph records.

(d) *The state of scientific knowledge* also affects the relative stability of technologies. For long years it seemed impossible to "break the sound barrier" in airborne vehicles. Yet once scientists broke the barrier, new instabilities quickly made their way into the technologies for manufacturing even subsonic planes.

Technology is correctly viewed as a universe because it is a system of its own whose field of influence is the entire globe. Major technological changes such as the miniaturization of computer circuits quickly spread throughout the world, even in places where no autonomous technological innovation takes place. Indeed in such locales a competitive edge based on technology can most easily be established. Transnational corporations (TNCs) also know intuitively that a competitive edge which has been lost or diluted in "mature" markets can be regained in less mature markets. The history of TNC investment attests to the profitability of technologies and derived products in Third World sites long after the competitive edge, or even basic marketability, has been lost in original industrial sites.

TNC marketing practices also suggest an interesting gloss on the basic theses of Latin American dependency theorists.[8] This added dimension may be called the "sequence of dependency." (Chapter 4 of this book discusses in greater detail the role of transnational corporations in technology transfers. At this point it suffices to mention the elements of the "dependency sequence.") The sequence is initiated when the dependency of purchasers is expressed in their need for a varying spectrum of goods provided by outside sellers. Initially, public and private firms in less-developed countries depend on outside suppliers for *capital.* This need leads them to offer inducements to direct investment and other forms of supplying capital, such as loans or grants. After pressing capital needs have been met, however, or at least mitigated, the most pressing demand felt in underdeveloped economies is to import *technology.* Once again, varied incentives are held out to those who can satisfy this demand.

But what can transnational corporations offer to poorer countries once the latter have met their needs for capital and for technology? Many firms whose capital or technology is no longer sought or welcomed are courted for their *managerial expertise.* But in one sense, managerial expertise is simply a particularly intangible kind of decisional technology, special in that it can be gained only after long years of experience.[9] Moreover, it is usually enterprise- or firm-specific (not industry-specific) or general.[10] Thus a firm lacking managerial expertise can acquire it only by an ongoing transfer process which must be contractually negotiated.

The final component—after capital, technology, and managerial expertise have been obtained (hypothetically, of course) by less-developed purchasers—is *access to markets*. Prerequisites of access are an existing network of contacts, specialized legal and bureaucratic skills, and rapid information-processing abilities without which final products would not move fast enough or far enough to amortize the high production-input costs of capital, technology, and managerial expertise. Again is illustrated how tightly technology is bound to the dynamics of competition. (To make this relation explicit is necessary because many writings treat technology as though it were some good transferable independently of competitive laws.)

The four-step dependency sequence just outlined grows in importance even as increasing numbers of Third World countries reduce their dependence on imported capital, because they remain dependent, nevertheless, on outside suppliers of technology. Venezuela is an illustrative case; now that the country is self-sufficient in capital, it has launched an ambitious program aimed at reaching a high degree of technological self-sufficiency.[11] Arab oil-producing countries and Iran likewise no longer need capital from industrial centers; yet they still need technology. And corporate sellers of technology are quick to understand that the locus of their present "competitive edge" may shift once again. For this reason they strive to transfer their technology in ways which disassociate it from their managerial expertise. And why? Because such expertise is the next asset down the line which assures competitiveness to its possessors.

Is the technological universe, therefore, blindly condemned to grow in present modes, or can technological maturation be reached within patterns of steady state?[12] The question is whether qualitative improvement can replace quantitative growth as the driving force of the evolutionary dynamism of technology.

Unlimited Growth or Steady State?

Business leaders throughout the world speak glowingly of the benefits of growth. A typical encomium appears in the annual report of one corporation in these words:

> Whether or not it is expressed in words, there is a philosophy that guides the destiny of every corporation. The philosophy under which Koppers operates consists of a number of tenets. One of those tenets emphasizes the need for growth.
>
> To the public eye, growth becomes visible through rises in

> sales and earnings. Underlying those statistical gains is a recognition that the corporation has been successful in fulfilling its primary mission: to upgrade resources in order to provide the abundance demanded by society in its efforts to improve the human condition. . . .
>
> Growth can provide the opportunity for new challenge and relief from routine.[13]

For the managers of Koppers, as for their peers in other companies, growth "improves the human condition." To questions regarding equity and social justice, they reply that distributing new benefits can simultaneously right old wrongs and satisfy new needs. This conventional wisdom also asserts that competition is legitimate because it powerfully stimulates growth. Nevertheless, competition can be dissociated from growth paradigms, and technology itself can be viewed as competitive beyond the confines of standard, purely quantitative images of growth.

Technology does not itself create or cause competition in arenas of exchange; on the contrary, it is competitive because the arena in which it circulates responds competitively to market stimuli. Thus, technology could conceivably cease being the source of a marketable "competitive edge" if the incentive systems governing exchanges were altered. Such a change does not necessarily imply making technology static, however, inasmuch as various other stimuli can propel or elicit technological improvements: status emulation, the desire to solve problems, the drive to know more, or the urge to improve the quality or to increase the durability of present tools. Implied here is the belief that *qualitative technological growth* is fully compatible with nonmonetary models of competition. Recent theorists use terms such as *steady state* for models of economic progress or stress the need for "organic" instead of disjointed growth.[14] Indeed technological maturation may prove more essential to the success of these efforts than it is to present growth models. The key and unavoidable questions remain, however: What is technology for? Which values, politically arbitrated and ethically confirmed, should command technological choices? Arbitration is necessary because, as Victor Ferkiss writes,

> now that technology has given us the power to destroy these life processes and to alter the nature of the human species, *every decision is intrinsically political.*[15]

Borremans and Illich make the same point and conclude that "what is necessary today is the political control of the technological characteristics of industrial products."[16]

Some government planners define technology's role in harmony with an "organic growth" model of development. In order to create a decision-making system which could integrate and balance social, economic, and environmental processes, they devise a conceptual guidance system for making budgetary and programmatic decisions. What is germane here is simply the recognition by planning teams that only a "highly technical discipline" can enable them to control and redirect growth toward humane ends.[17]

The need to innovate qualitatively is salient in diverse approaches to "appropriate" or "intermediate" technologies. The priority goal sought by all is sound human development which shatters both the market determinism of capitalist growth and the rigidities of centralized, socialist planning. Serious advocates of "steady state," "organic," or "human scale" development acknowledge that their own goals cannot be reached without technology—hence their quest for alternative models of technological maturation, placing special emphasis on "self-help technology" which aims primarily at helping the rural poor develop their own economies. Criteria for self-help technologies are labor intensity, low cost, maximum utilization of local materials and skills, the protection of resources and environment, and easily managed scales of operations. E.F. Schumacher judges that "donor countries and agencies do not at present possess the necessary organized knowledge of adapted technologies and communications to be able to assist effectively in rural development on the scale required."[18] Were "adapted technologies" available, therefore, they would enjoy a "competitive edge" in meeting important unsatisfied needs not presently met. The experience of Schumacher's Intermediate Technology Group[19] attests to the need for new and better (but cheaper and simpler) technologies in the Third World, particularly in food-growing, water-harnessing, machinery design, health services, and housing-construction.

Most discussions of alternative technologies—called, variously, *radical, soft, intermediate,* or *appropriate* technologies—center on rural questions. Nevertheless, they raise issues germane to urban living and to industry, in short, to "developed" countries. This relevance is emphasized by the Community Technology Group in Washington, D.C.[20] Even US city-dwellers, argues the group's founder, Karl Hess, need to develop high degrees of self-sufficiency and achieve mastery over small-scale technologies.

My argument can be summarized in a series of related propositions and questions:

(1) *Technological expansion, as presently conducted, is highly wasteful of resources. If, therefore, resource and ecosphere conservation become priority goals, should technology be allowed to keep expanding?*

The answer is yes, provided that expansion takes place in a different mode. Conscious efforts need to be made to achieve qualitative maturation of technologies overtly designed to assure ecological integrity, more manageable scale, and greater accessibility to poor people.

(2) *If untrammeled wasteful growth is undesirable, is stagnation the only alternative?*

No. In "organic" or "steady state" growth models, quantitative gains are not eliminated but subordinated to qualitative improvements, to the mode in which the growth is realized, and to considerations of social costs paid to achieve it. Growth, in short, is sought in ways which foster a cluster of stipulated values.

(3) *Can technological evolution adapt itself to the requirements of such growth?*

Yes, on condition that the basic value options, development strategies, and technological-development policy of a society are clearly defined and coherently pursued. (The "vital nexus" among the three is discussed at length in later pages.)

(4) *What other changes must occur before such an altered course in the direction of technological evolution can become possible?*

Many prior changes are required. First, widespread value transformation must wean people away from their infatuation with mass consumption and endlessly wasteful changes in design, shape, packaging, and color. The vision of a good life [21] must center on "sufficiency for all" defined in a dynamic way which balances quantitative growth against other values. Consequently, broad political agreement needs to be reached as to the desirability of placing some ceiling on the scale of technologies and on kinds of production. Furthermore, the education of engineers, designers, and planners must be revolutionized to release them from their servitude to the technological imperative.[22] This may clearly be the most difficult task of all.

At the conclusion of this work I shall return to these issues and discuss technology assessment and the revitalization of culture in the face of technology's standardizing influences. I must first, however, examine a relationship which both explains and obscures the dynamism peculiar to technology as a social system. This is the "vital nexus" which links any society's basic value choices to its preferred development strategies and to its attitude toward technology expressed in policy.

The Vital Nexus: Value Choices, Development Strategy, Technology Policy

As stated above, technology is both a system of its own and a component of larger social systems. One must, accordingly, analyze its workings by alternatively probing technology's inner dynamics and its links to broader social processes. It is particularly useful to analyze the link which binds society's basic value options to its preferred development strategies and to its technology policy. This "vital nexus" of the three is well illustrated in the case of the world's largest poor country, the People's Republic of China.

Mainland China openly affirms the central importance of value transformation on its road to development.[23] And a growing body of scholarly literature is now available describing China's approach to technology.[24] It lies beyond the scope of the present book to analyze or even to summarize China's technology policy. So as to illustrate the importance of the "vital nexus," however, it is worth recalling the importance attached by the Chinese to coherence among basic value options of their society, their road to development, and their technology policy. Huge problems faced China when Mao acceded to power in 1948: society had to be reconstructed from the ruins of war and foreign occupation and mobilized, along new ideological lines, to produce more abundantly and efficiently. Countless institutional problems identical to those faced by other nations in quest of "development" had to be solved. Among these were the creation of a universal educational system founded on social merit and participation instead of on hierarchy and privilege, the provision of health services to a population which remained largely rural and suspicious of "Western" medicine, and gaining effective access to foreign technologies. By all accounts, China's monumental efforts in these domains have brought relative success (whatever be one's final judgment as to the social and political "costs" incurred). Of special interest, however, is the explanation offered by the Chinese themselves as constituting the key to success.[25] One must, say the Chinese, center efforts on overall incentive systems operative in society and base these on values consonant with revolutionary objectives. One common formulation of the approach reads: "Values command politics, politics commands economics, and economics commands technique." Central values adopted and disseminated are:

(a) *the need to acquire "revolutionary consciousness"*

Developing this consciousness requires a new reading of China's historical past, which explains the causes of its subjection to foreigners and the perpetuation of indigenous privileged classes. This study also highlights the historical potential the nation presently possesses for creating a new society now and in the future.

(b) *a vision of "austerity" as preferable to a model of affluence*

Austerity is here understood to mean "sufficiency for all" obtained by "strenuous striving" to increase production and productivity. In pursuit of that sufficiency, all must make optimum use of every resource and struggle mightily, not only against the acquisitive spirit but also against "alienating" oneself in the desire for future goods. One pedagogical theme repeatedly stressed is the primacy of moral, over material, incentives. This primacy, it is stated, is the pillar upon which must be built the edifice of solidarity and the "serve the people" ethic. Austerity, therefore, is viewed not as a necessary evil to be tolerated in times of scarcity or initial poverty but as a permanent component of authentic socialist humanism. The assumption underlying this belief is that people are as deeply "infected" by the virus of acquisitiveness in their desire for future goods as they are by clinging to goods already possessed.

(c) *a commitment to high degrees of equality and participation*

The endless struggle against differential expropriation of benefits and against elitism are a social and institutional expression of this value choice.

(d) *a strong affirmation that the single most important resource for development is human will, collectively and responsibly mobilized*

This insistence leads to an attitude of thinking that no problem is insoluble, even in the absence of what are considered to be "normal" resource requirements of a material or a technological sort.

If, therefore, a society were to make these value choices (although no society can perfectly, or with full consistency, practice them!) and if, furthermore, it were to try overtly to formulate a "road to development" (or a development strategy) which coherently promotes these values, then obviously different criteria for policy will emerge than would otherwise be the case. It would become essential, for example, to decentralize productive investment, to institutionalize maximum self-reliance in local units, to combat tendencies which create or perpetuate chasms between intellectual and manual labor, etcetera.

The choice of initial values also has its impact on the precise formulation of technology policies. There is no need to review Chinese technology policy in detail here; nevertheless, one notes that great care is taken to allow at important sectors of production (the manufacture of consumer goods, agricultural production, and the provision of basic services) for grassroots technological innovation and shared research responsibility.

Although a worthy example of how the "vital nexus" may work, China is not perfect; it is no social paradise but an historical experience fraught with contradiction. Yet few societies strive so mightily and so explicitly to design development strategies and technology policies in accord with prior value choices. What is more, few nations attempt to formulate these choices so clearly and so vigorously.

My contention is that a direct correlation exists between the degree of linkage among the three component elements of the "vital nexus" and the quality of technology policy itself. Thus one can frame satisfactory technology policies—on the international, regional, and national levels—only to the extent that one is clear and firm regarding basic social values and development strategies consistent with these values. Many country planners and politicians, it is true, do articulate goals, albeit in purely rhetorical fashion (emphasizing, let us say, "developmental equity" or "relative technological autonomy"), and yet they refrain from adopting the strategies and policies which would render these goals feasible. In contrast, what China's example brings into sharp focus is an important lesson in how development can be guided by values and how value transformation can indeed become the main road to development.

In an earlier work I have argued that development decisions are not primarily economic, technical, or even political in nature.[26] Rather they are moral options around three vital issues: the criteria of the good life (the relations between the "fullness of good" and the abundance of "goods"), the basis for just relations in society, and the principles for adopting a proper stance toward the forces of nature and of that "second nature" we call technology. What renders these choices specifically developmental is the modern setting, characterized by the massive scale of operations; technical complexity and its attendant division of labor; multiple interdependencies which bind each part to the whole and the whole to each part; and the ever-narrowing time lag between the impingement of social changes proposed or imposed and the responses societies must make to assure survival, identity, or creative assimilation of change. Hence the development strategy any nation adopts and, *a fortiori,* its policy in more limited domains such as technology are necessarily linked to its value options.

Specialists usually discuss development strategies in terms of relative priorities: investment in industry over agriculture, in human resources over infrastructure, tax incentives to foreign firms over increased credit to native firms, and so on. Although planners rarely advert explicitly to the nexus between values and strategic priorities, its existence is undeniable. Thus if one adheres to the value of greater egalitarianism, one will tend to favor improvements in agriculture over industry, small technology over mass-scale techniques, subsidies to local firms over tax holidays to transnational corporations, and popular decision-making over exclusive reliance on experts. The same interdependence between strategy and values exists at the level of ideology. If one chooses capitalism, with its implied effort to integrate into the world market, values such as self-reliance and local innovation are relegated to the background. If, conversely, one adopts a communitarian socialist strategy of development, one will prefer gains in economic independence to pure efficiency and one will attach greater weight to social justice (in the land-tenure system, for example) than increased output. In a word, *value choices and development strategies are tightly linked.* When one introduces a third element into the equation, namely, technology policy, the nexus tightens still more.

What is to guide technology policy if not basic values and the strategies derived therefrom? Surely not mere considerations of technical self-sufficiency, uncritical aspirations for technological modernity, or imitation of technological pioneers. If technology policy is to have the consistency of sound decision-making, it should flow from the basic value choices underlying the selected development strategy.

Many national-technology policy-makers appear oblivious to this link; yet no technology policy can succeed if it is not expressly designed to reinforce the social values pursued, in some scale of priorities, by the development strategy adopted. Certain approaches to technology are evidently more congenial than others to this unity. To illustrate, if Tanzania's commitment to self-reliant development which builds on the communal values of its largely rural communities is a serious objective, one would expect its technology policy to assign a wide role to "soft" technologies aimed at increasing productivity through optimal use of local resources. Or, as one reflects on Algeria's declared goal of achieving the full range of industrial capacity for internal and export markets, different technological measures recommend themselves. Among these are: importing foreign technologies to build up competitive capabilities, training nationals in order to limit dependency to the briefest possible period, and achieving a coordinated bargaining posture so as to avoid outside exploitation.

To affirm the existence of a "vital nexus" among value choices, development strategy, and technology policy is not to state that all technology policy-makers in the Third World derive their policy from the other two elements. But it does mean that they *should* do so if they are to avoid two dangers. The first is falling into contradictions between basic development goals and technological choices. The second is becoming imprisoned by the greater or lesser degrees of determinisms which are inherent in technology itself or which flow from the uncritical acceptance of conventional technology transfers. Therefore, the best way to design policy is to advert explicitly to the "vital nexus"; it cannot be ignored with impunity.

* * *

Part One of this work has described the technological universe. Technology has been called a two-edged sword because it is ambivalent, promoting certain values while threatening others and creating new servitudes as it frees its users from old constraints. Because the technological universe is not static, I have also described its dynamics, focusing on technology as a kind of artificial nature constantly evolving at a quicker pace and with greater unpredictability than nature itself. This mutability of technology was then placed in the context of economic competition, a major stimulus to social change. Also delineated was a "sequence of dependency" which less-developed nations might envisage breaking by progressively reducing their reliance on outsiders for capital at a first stage, then technology, later managerial expertise, and finally, access to markets. A further dimension of technological dynamism is its vital linkage to broad value choices constantly being made within changing societies and to preferred development strategies.

One central premise of this book is that value conflicts in international technology transfers are traceable to two distinct sources: the value ambiguities of technology itself (even in its matrix of origin) and the specific channels and mechanisms by which technology flows from rich to poor countries. The first of these has been examined in Part One. It is now time to examine how technology is transferred from "developed" to "less-developed" countries (LDCs). This exercise is conducted not for its own sake but to shed light on one crucial question: Do such transfers impede or aid genuine development for all? This question is now addressed in Part Two.

Part Two:

TECHNOLOGY TRANSFERS: AIDS OR OBSTACLES TO DEVELOPMENT?

Introduction

To speak of "transfers" can be misleading because technology is usually bought and sold internationally in a predominantly seller's market. Hence some experts prefer to speak of the "commercialization" of technology. Yet this term is too exclusive because many technological exchanges are subsidized and take place outside commercial circuits. It is more accurate to speak of the "circulation" of technology, a term which embraces all forms of transfers—commercial, free, and subsidized. Nonetheless, as a concession to the vast literature on "technology transfer," I shall use the term in the broad sense just given to "circulation."

The central question is whether, on balance, technology transfers aid genuine development or impede it. One cannot determine if technology transfers are beneficial or harmful to development without first analyzing how such transfers take place. Chapter Three, accordingly, examines the mechanisms and channels of these transfers.

3/Mechanisms and Channels of Technology Transfer

Technology does not exist in a vacuum: it is embodied in products, processes, and persons. Similarly, transfers are made via a wide array of concrete channels, both institutional and procedural. The institutions through which technology circulates are here called *channels*, whereas the instruments employed are the *mechanisms* of technology transfer. To some degree this distinction is arbitrary because most institutions use identical mechanisms. Moreover, the instrumentality favored by a given institution in transfers is often distinguishable from the institution itself only in the observer's mind, not in workaday practice. Yet for clarity's sake it remains useful to keep the two distinct.

Channels

Among the leading channels or institutional vehicles of technology transfer are transnational corporations, think tanks, foundations, professional associations, academies of science, universities, labor unions, voluntary agencies, individuals, and public agencies of all types, including national governments and international agencies. Not all are equally important, nor do technology transfers command a proportionally equal amount of their energies. The further away one gets from technologies embodied in products and processes, the more likely is one to be engaging in less tangible forms of transfer, or even in what is not strictly a transfer but technological education to counterparts. Most of the following pages center on technology transfers which are product- and process-embodied. The sole exception is the case of person-embodied "decisional technology": know-how for diagnosing complex problems and formulating choice strategies for site, scale, and level of technology. Decisional technologies are vital because, in the words of Argentine physicist Jorge Sabato,

> the ability to conduct a feasibility study with its own means is the single most revealing touchstone indicating when a country has reached an acceptable level of technological autonomy.[1]

These comments suggest the central role played by consultant firms, a role no less important than that played by large manufacturing or extractive (mining) firms. Indeed the competitive edge they enjoy is largely conditioned by the degree of linkage these firms maintain with consultants in such fields as finance, engineering, design, conjunctural studies, and marketing. Although many manufacturing firms possess their own consulting capabilities in key sectors, outside consultants are usually quicker than they to detect shifts in LDC government policies and more flexible in translating such intelligence into operational terms. A brief sketch of how consultants work will, therefore, be given in later pages. But because their grip on technology is less tangible than that of manufacturing or extractive firms, it is advisable for pedagogical reasons to examine first how firms of this type carry on technology transfers.

Manufacturing, Extractive, and Service Firms

A wide array of instrumentalities is used by manufacturing, extractive, and service firms to transfer technology. The burgeoning literature on the subject lists them: direct investment; exports of machinery, equipment, and products; industrial and trade fairs; licensing contracts of all types; training arrangements of various sorts; supervision or quality control at production sites; and technological conferences, seminars, and workshops.[2]

Recent debates over TNCs have provided a clearer view of the diverse benefits attendant upon technology transfers. For a long time governments, national firms, and scientists and technicians in LDCs erroneously presumed that technology was being transferred for their benefit under all the mechanisms listed above—hence their largely uncritical endorsement of public relations releases issued by TNCs regarding the benefits of transfers. Once made critically aware, however, of the difference between mere geographical or intrafirm transfer and genuine indigenous assimilation of technology (with mastery, control, and improved ability to gain future autonomy), LDC governments began pressing for new terms in transfers. At first TNCs denied or minimized the problem. Once subjected to sharp critique, however, they realized that they could no longer proceed in this way. Accordingly, more enlightened firms are now preparing themselves to renegotiate terms on the basis of mutual concessions, the extent of which remains unspecified. One reason that TNCs are slow to make real, instead of merely rhetorical, concessions is that their interests are well served by conventional modes of "technology transfer": direct investment, intrafirm transfers, intracompany transactions with affiliates and subsidiaries, licensing contracts with client firms. A brief review of these preferred modes explains why companies favor them.

Direct Investment

Direct investment is easy to manage. Once a TNC acquires the expertise to handle international currency transactions, to recruit personnel from various cultures, and to master the logistics of transnational transportation, the decisional and operational procedures required in direct investment are congenial to the basic values of corporations. Complex negotiations with host governments are doubtlessly troublesome, as are red tape for import licenses and profit remittances. But these are mere procedures easily learned: *they pose no threat to the basic interests or work styles of corporations.* When large corporations engage in direct foreign investment, their conduct of technology transfer overseas differs little from the domestic in-house communication of technological findings. Engineers, chemists, systems analysts, and quality-control specialists routinely "plug in" the results of R&D and of innovations in their fields to ongoing firm operations, giving little thought to such notions as competitive edge, market advantage, cost efficiency, and the coordination of overall procedures. When they move to a subsidiary or affiliate overseas, they operate in identical fashion. Host governments, however, are more sensitive than home governments about such issues as job-creation, national control over technology, and losing specialists in the brain drain. For the TNC these are disquieting but basically irrelevant externalities. When these concerns create serious public relations or "image" problems, or when host governments undertake to change the ground rules of negotiations, TNCs react. Their problem is simply to assess whether they have more to gain or to lose from making concessions. Concessions are, of course, portrayed to the public as moves engendered by the "cooperative" spirit of the enterprise, and realistic government officials who understand these practices are prepared, once they have obtained their substantive concessions, to behave like all experienced negotiators and "let the other side save face."

Pressured by new demands from governments, many TNCs which have favored direct foreign investments only when they could be sole owners of enterprises are now agreeing to become minority equity holders in joint ventures. To illustrate, the Cabot Corporation, one of the world's largest manufacturers of carbon black, has acceded to joint ventures, notwithstanding its earlier policy of sole ownership within Third World countries. The company settled for 50% ownership in Malaysia and Iran, and in Brazil it sought an equity share lower than one half so as to be legally able to charge technical service fees to its Brazilian affiliate.[3] One lesson to be learned from recent trends is that although government restrictions in LDCs may inhibit the freedom of companies to operate as they had in the past, these companies often still have more to gain, on balance, from entering into joint ventures than from avoiding them completely.

Once TNCs accept such partnerships, their preferred modes of transferring technology are adjusted accordingly. Their LDC partners usually demand greater dissemination of technology from the head office to the subsidiary venture. And specific clauses may even require the head office to train not only officials of the joint-venture firm itself but others whom the host government wishes to instruct in certain technologies. Why do transnational firms accept such constraints? Often it is because they can do no better in a given market; in other cases, it is to assure a privileged position in selling raw or semiprocessed materials or equipment, or to gain information about local markets or supply sources, or to prevent a competitor from gaining the same advantages. The more LDC governments and firms understand the goals of TNCs, the better prepared they are to negotiate satisfactory technology transfers with them. In the absence of pressure upon them, TNCs will transfer technology in ways which are easiest and least disadvantageous to themselves. But to avoid losing other advantages, TNCs will make concessions on modalities of technology transfer and on legal terms of direct investment.

Licensing

Interviews with transnational executives reveal that overseas licensing is, for most of them, a relatively minor source of income. Published data are scarce.[4] Yet there is no doubt that such licensing is advantageous to TNCs under many circumstances. The decision to license overseas may be dictated by such considerations as the needs to:

(a) obtain supplementary earnings from technologies whose period of competitive advantage in primary home markets is drawing to a close

(b) gain access to markets where direct investment is excluded, either by formal policy, general practice, or specific discrimination

(c) seize opportunities to improve a technology in special circumstances which approximate those found in third-country markets of the parent firm

(d) gain the goodwill of selected governments by supplying them with technology even if the economic advantages of doing so are not great (seeking to maintain "friendly" relations which can be useful in other domains)

(e) obtain side benefits in the form of favorable corporate publicity

Licensees, in turn, enter contracts for varied motives but usually because they lack the technology in question, cannot produce it themselves (at least quickly or inexpensively), and need it to process their goods. In other cases, licensees are driven by the desire to process up-to-date technology used by leading firms (usually in "developed" countries) in a given industry. Considerations of scale also weigh heavily on licensees; if their markets are too small to justify manufacturing the equipment needed or even investing in R&D to generate process technologies, they buy it from outside suppliers under licenses —the "normal" way of getting technology across national or enterprise borders.

Basic terms of licensing agreements vary widely within companies, industries, and countries. Some differences are traceable to the nature of technologies as product- or process-centered. For illustrative purposes it is worth noting that typical *product-embodied* technology licenses contain such clauses as the following:

(a) Exclusive rights to sell the product are given to licensees within stipulated territories. Licensors, however, are sometimes allowed to sell their products in that same geographical area.

(b) Advertising of the licensed product—usually bearing a well-known trademark or trade name—is strictly controlled.

(c) Licensees must supply random samples of their products to the licensor, who controls quality, a measure considered essential to preserve the "good name" of the primary producer. Detailed reporting is also required from licensees on: sales-promotion efforts, the qualifications of personnel assigned to licensed production, the licensee's evaluation of competitors's products, etcetera.

(d) Licensors usually claim rights to inspect a licensee's factory and laboratories.

(e) Licensors pay licensees for material samples sent for quality inspections.

(f) Licensors require partners to register products in all countries which require such registration.

(g) Licensees are usually obliged, under threat of losing exclusive rights, to place a licensed product on the market within a specific time period (often twelve to eighteen months). Grace periods or extensions can be negotiated, however.

(h) Royalty percentages and modes of payment are spelled out in detail. For example, in one specific case royalties were 5% on the first $1 million of sales, 4½% on the second $1 million, and 4% on sales beyond $2 million. Where host-country legislation (as in Brazil) fixes a ceiling lower than 5%, sales in that country are not included in the accounting of gross sales.

(i) Technological know-how must be treated by licensees as confidential knowledge, and violation of confidentiality (within specified time limits) can lead to contract cancellation.

(j) The duration of licensing contracts varies widely. Often it is three, five, or seven years; occasionally, renewable.

Worthy of note in most contracts is the subordination of monetary clauses to clauses assuring licensors control over their own market areas and those assigned to licensees. The greater concern for control illustrates a general relationship found in most commercial technology transfers: that whereas host-country governments view the acquisition of technology as an end in itself, TNCs and their client or affiliate firms in LDCs see technology transfers as simple means for assuring success in their marketing operations. Stated another way, TNCs are interested in technology transfers to the extent that these foster their own purposes; hence they often view limitations imposed by LDC governments as unreasonable. Conversely, LDC governments often seek to disassociate technology transfers from their role as instrumentalities of corporate activities; the LDC governments' own aims are to build up their pools of technologically skilled nationals, to circulate foreign technologies among many units of national industry, and to minimize hard-currency flows outside the country. Moreover, LDC host governments rarely approve of contractual restrictions on sales to third countries.

Many conflicts in licensing contracts arise from a discrepancy between primary and secondary benefits anticipated by the respective parties. Indeed most licenses are signed with a view to achieving multiple objectives. But Objective Number 1 for the licensee (to obtain a modern and competitive technology, let us say) can be contrary to Objective Number 2 for the licensor (to gain a foothold in a difficult market). Consequently, criteria used by suppliers and utilizers of technology to determine whether the transfer is effective often vary. Mutual dissatisfaction occurs as frequently as do compromises dictated by necessity. Nevertheless, licensing contracts continue to be signed in large numbers because LDC firms desire technology and TNCs seek access to LDC markets. Mutual advantage is still possible, of course, even where priorities are divergent. But it is evident that minimum satisfaction obtained by each partner in its primary objective is the bottom threshold below which contracts will be viewed as exploitative.

Parallel tensions exist in technology transfer contracts between LDC host governments and nonprofit agencies (universities, philanthropic agencies, religious institutions, etcetera) based in developed countries. The rich-country university is often eager to build up its own institutional capacity to teach and conduct research on specific problems. Thus it treats LDCs as a testing ground for its methodologies. Accordingly, it will contract, often with third-party funding (from, for example, the Agency for International Development, the World Bank, or the United Nations Development Programme), to transfer technology to some LDC host agency. The latter, however, may assign priority to training local experts, to solving specific problems, and to mastering the methodology at stake. These two priority scales are not automatically incompatible; nevertheless, the primary objectives of both parties must be minimally satisfied if they are to rest content with arrangements. LDC governments are now pressuring TNCs for more favorable terms in technological licensing precisely to assure that their own priority objectives as well as those of LDC firms will be duly protected.[5]

A close link exists between licensing as a mode of technology transfer and contractual training agreements. Many licensees are more interested in the license clauses related to training than with those bearing exclusively on the transfer of products or processes.

Especially in process technologies the right to send LDC personnel to rich-country factories and laboratories is a vital feature of technology transfer. Training, besides enhancing a firm's capacity for fruitful assimilation of technology, is also a valuable step toward eventually acquiring the ability to produce one's own technology. The same objective is served by those licensing contracts wherein supplier firms provide updating courses and technological initiation seminars to personnel of licensees. No doubt many nonproprietary technologies circulate through scientific publications, open conferences, and industrial fairs. Yet the ability of LDC governments or firms to benefit from these exchanges is often conditioned by the existence, within them, of what Thomas Allen calls the "technological gatekeepers."[6] "Gatekeepers" are those persons within a firm—engineers, technicians, laboratory workers, other researchers—who "keep up" with professional and scientific journals and maintain ongoing contact with the technological community outside their own firms and industry. Contact is maintained through regular attendance at foreign scientific and professional society conferences and sustained correspondence. Effective gatekeepers are well integrated in two networks: an external network of foreign information sources and an internal network of domestic colleagues within the firm to whom the required information can be transferred for practical application. Unless each network is used frequently and is diversified, it will atrophy. One policy recommendation implicit in Allen's studies is that firms in less-developed countries should deliberately structure the operations of their technological gatekeepers.[7] Such an innovation comprises but one dimension of tightening the "triangle" among policy-makers, production units, and researchers.

Industrial fairs constitute an indirect, albeit important, mechanism of technology transfer. According to recent reports China has found ingenious ways of using such fairs to high advantage.[8] The Chinese negotiate fairs of long duration and systematically organize visits to exhibits by engineers, students, and others who can assimilate the technology on display. Upon termination of the fair they also purchase equipment at reduced prices without, however, signing technological training or maintenance contracts. These exhibitions possess, in Chinese eyes, an educational rather than a commercial character. Nevertheless, in 1972 China began importing larger quantities of advanced industrial technology. Since then it has had to increase the number of foreign technicians it admits inside its factories.[9] China is mentioned here only to illustrate an important general point, namely, *the flexibility with which conventional mechanisms of technology transfer can be used to yield higher assimilation of technology when the clear political will and organizational ability to do so exist.* This principle can be applied to such routine modes of dealing with suppliers of technology as: the purchase of machinery and equipment, industrial fairs, distribution of samples or prototypes, licensing contracts, training arrangements of various types, the staging of conferences and seminars, data banks for centralizing available nonproprietary technological information, and consulting contracts.

At the plant level no less than at the national planning level, technology is "transferred" most successfully when the final utilizer either expresses a need for it or has a voice in defining the problem to be solved by that technology. Jack Ruina, professor of electrical engineering at the Massachusetts Institute of Technology, argues that the first condition for successful technology transfer is that the "recipient" already know something of the whole process and be able to take further steps to see precise connections among all elements of the technological process.[10] If, he adds, the final consumer of technology plays a role in formulating the initial problem and asking the prior research questions, the distance between the problem as conceptualized in an R&D laboratory and perceived in the daily world of problem-solving is greatly reduced.

This lesson is not lost on those engaged in day-to-day industrial transfers who try to create conditions which maximize their ability to absorb transferred technology more creatively. Ultimately, however, this maximization would require endowing LDC nations with independent research capacities, an aspiration thwarted up to now by the developed countries' almost total monopoly of industrial R&D.

The R&D Monopoly

Some 98% of the research and development expenditures of nonsocialist countries are made in rich countries, only 2% in developing countries.[11] Policy-makers in poorer countries, therefore, view with alarm the monopoly of research. Transnational corporations, in turn, defend their near-monopoly on grounds that only in advanced countries do conditions exist which favor success: the availability of large aggregates of capital, a pool of skilled researchers, the proximity to primary manufacturing and marketing units which makes R&D responsive to practical constraints, and a supportive attitude toward R&D in society at large. Scale is again a central question; many research managers interviewed declared that decentralized R&D investment in less-developed countries is impossible because production scales do not permit amortization of high and risky research costs. In some circumstances it might be profitable to consider building R&D units in Third World countries. But these favorable circumstances tend to be quite special, as illustrated in the following example.

Officers of Alcan Aluminum in Latin America declared in interviews conducted in late 1974 that the company is considering a large R&D unit in Brazil, because legislative restrictions imposed by the Brazilian government have made it imperative that the company find alternative ways to obtain the technology it needs for its Brazilian plants. (Specifically, these restrictions place ceilings on payments for technology imported from outside the country and limit sources of such technology to those not covered by national suppliers.) The reasons invoked to justify this consideration are the following: (1) Brazil can support an integrated aluminum/bauxite operation—extraction, smelting, processing, manufacturing—with full coverage of vertical integration activities; and (2) Brazilian plants sell to third countries as well as to the national market, which is significant because the Brazilian market, although larger than most Third World internal markets, would not suffice to warrant the projected investment were it not for anticipated export markets.

Company officials hastened to explain that few countries could satisfy the two conditions just described. They emphasize that in most cases TNCs, including themselves, do not wish to invest in decentralized R&D, especially in plants located in Third World countries. Given this attitude in corporation executives, the constraints facing LDC policy-makers eager to endow their countries with national R&D capacity take on special importance. Relatively few data are available as to the precise conditions under which transnational corporate managers decide to invest in permanent R&D facilities away from their home country.[12] Already it is apparent, however, that much R&D activity conducted by TNCs abroad, particularly in the Third World, is of short duration or is tied to fixed-term contracts with existing local research facilities. The general motives which lead corporate managers to make R&D investments are the desires to:

- facilitate and speed up technology transfer from domestic laboratories to foreign subsidiaries
- monitor local demand or supply sources for opportunities which escape notice in home countries
- increase chances of successful innovation by permitting the development of foreign innovation opportunities close to their market source

Ronstadt explains that apart from these motives, often shared by LDC host governments, another factor is at play: the desire of LDCs to retain their surplus of skilled scientists, engineers, or technicians who cannot be employed to the full level of their talents in purely local firms.[13] On balance, however, few efforts have been made by US-based transnationals to invest in R&D overseas other than in Canada or Western Europe.

Two directions for a reversal of trends appear possible. The first is bold action by individual firms. To illustrate, Bagó Pharmaceuticals, a medium-sized Argentine company, made a policy decision some years ago to achieve autonomy in R&D capacity.[14] One main reason for the decision was financial: multiple licensing costs were very high. Also present was the explicit desire to reduce the company's dependency on outside suppliers and to challenge the conventional wisdom which held that Argentine firms could never perform as well as foreign competitors. Additional considerations included the company's wish to gain the loyalty of a corps of researchers who would stay with the firm for long years, an attitude rarely encountered among expatriate researchers. Success was achieved within six years, one expression of which came when Bagő's researchers made technological breakthroughs which their former licensors then wanted to produce as licensees. A small number of other firms in LDCs have experienced similar success, usually by their own efforts and with no governmental subsidies.

A second means of breaking the R&D monopoly of TNCs is determined governmental action utilizing direct subsidies when necessary. Rhetorically at least, many LDC governments seek to have their own technological research infrastructure, particularly in branches of industry vital to exercising control over their national economies. Chile wants to gain autonomy in copper; Bolivia, in tin; Argentina, in meat and wool. Success is impossible, however, in the absence of a clear policy vigorously pursued and subsidized. The decision reached by the Argentine government in 1973 in the sector of nuclear energy for peaceful purposes is one example.[15] Notwithstanding the contrary consensus of world experts, the government rejected the option of enriched uranium (at that time obtainable only from the United States or the Soviet Union) to seek self-sufficiency in nuclear-energy production using natural uranium. One major obstacle stood in the way: permanent research capacity had to be created. It was created. Although these examples are of limited applicability, they do show that important gains can be won.

Since the fruitful assimilation of imported technologies is conditioned by the level of one's absorptive structure, the possession of native R&D capacity improves one's ability to make optimum use of foreign technologies. A coherent policy aimed at developing local R&D would also provide material, social, and moral incentives to entice research professionals to remain in their countries. Incentives should be institutional and not solely personal. Thus a government will invest in improving the quality of medical research and service facilities in small, remote areas while providing other incentives to doctors, nurses, aides, and paraprofessionals to apply their skills away from congested capital cities. But it is wasteful to endow a country with R&D infrastructure unless that research effort is effectively coordinated with industry's demands for technology. At stake is the circulation which must take place within the triangle—policymakers, industry, and research producers—articulated by Latin American specialists.[16] Little good comes from increasing the potential or actual supply of national technology unless that technology can be matched up with the national demand structure which habitually looks outside national boundaries to meet its technology needs.

(International suppliers have obviously had enough influence on internal demand structures to shape their size and form in accord with their own interests.) Something more basic is needed, therefore, than mere "transfers of technology" to national research institutes. Many such transfers are little more than uncoordinated educational programs to train or upgrade local scientific personnel. Whatever be their intrinsic merits, however, these programs often bear no relation to the production process. Consequently, efforts to build up an R&D infrastructure in LDCs must center on locating research within industry or in close symbiosis with it. To do otherwise is to waste funds and to court failure.

There is no compelling reason why LDC governments could not require selected transnational corporations to invest in local R&D as a precondition for operating within their borders or even for engaging in conventional technology transfers. Similarly, LDC governments could offer fiscal and monetary incentives to firms willing to build, set up, and supervise local research and development facilities. Because most international firms derive their competitive advantages from a variety of sources, of which technological advance is but one, it must not be assumed *a priori* that they will necessarily deem such demands to be intolerable, at least in industries characterized by relatively stable technology. One prevalent fear is that technologies adapted to Third World local conditions will prove unable to compete with those designed for conditions in the rich world. But much relativity attaches to this line of reasoning: rich countries themselves, under the pressure of long-term inflation and growing scarcities of food and fuel, may be obliged to shift to factor proportions in production closely paralleling those now current in underdeveloped lands. They too may be forced to maximize employment, to spare scarce capital, and to use technologies which do not deplete natural resources or spoil the environment and which can work on smaller scales. Many years must surely elapse before Third World countries, taken as a whole, can build up R&D infrastructures of a size, diversity, and strength comparable to what now exists in rich nations. But for this very reason, TNCs and international consultant firms could view the implantation of such facilities in the Third World as a profitable enterprise for long periods to come. TNCs with a long R&D tradition need not necessarily feel threatened by the coexistence, alongside their own home-country R&D facilities, of local Third World R&D units which will still have to compete with them to obtain skilled personnel, sophisticated laboratory equipment, and access to the pool of general knowledge germane to the operations envisaged. Clearly the mere physical presence of research laboratories in less-developed countries will not guarantee that technology developed and adopted will be congenial to local needs any more than the implantation of manufacturing plants will, of itself, assure a type of production geared to meeting priority needs of the majority of local populations. The answer lies in a national policy aimed at shifting the direction, composition, and quality of research toward social purposes broader than those pursued by corporate R&D units. If such energetic action is lacking, the "indigenization" of R&D will prove as disappointing to Third World developers seeking to reduce dependency as did import substitution in the industrialization phase.

Many findings generated by R&D laboratories in rich countries reach Third World sites only through the mediation of consultant firms which are expressly set up to keep abreast of research and to help spread it. The important role such consultants play in technology transfers must now be examined briefly.

Consultant Firms[17]

The term "consultant" when applied to firms is generic; variations within the genus are wide. Some consultants provide design and problem-solving services in limited domains—hydraulic studies, food-processing technology, or refrigerated transportation. Others undertake virtually any task, from evaluating the managerial efficiency of a company or government, to making feasibility, site, or design studies of a paper mill; to advising churches how to invest their portfolios in an "ethical" manner. The capabilities of consultants include the ability to devise equitable tax systems, plan the reform of public bureaucracies, install modern systems of data-gathering and -processing, formulate national development plans, map out regional tourist policies, or train managerial decision-makers. Sometimes consultant firms are specialized interpreters of information whose role is to organize all circumstantial knowledge which potential investors, governments, labor unions, or other economic actors might conceivably find useful. Large consultant operations of this type operate like a consulate general: they post overseas their equivalents of labor attachés, economic advisers, political analysts, legal experts, statisticians, and public-information specialists.

It is no idle question to ask why consultants are necessary. Answers given by consultant firms themselves are revealing. In a folder prepared for potential clients, Arthur D. Little, Inc. states that:

> ADL considers its principal business to be the management of change and the optimal blending of change with continuity.... Each time we undertake a new assignment, we assemble a new case team representing whatever skills are needed to understand all aspects of the problem and their interrelationships. What this means...is an approach to problem solving which combines:
> —General expertise in all the change areas that affect the... field—government regulations, technology, economics, society;
> —Specific expertise in marketing, organizational development, strategic planning, forecasting, modeling, finance; and

—Most specifically, skills in operations, research, and planning for specific situations within the broad context of a burgeoning industry.[18]

Another prestigious world consultant, Business International Corporation, declares itself capable of:

providing fast, reliable information needed for corporate planning and decision making
alerting corporate management at home and abroad to new opportunities and dangers
discovering, explaining and interpreting new international management techniques that will advance profitable corporate and economic growth
analyzing governmental measures that will make for sound economic growth and greater international cooperation and that will pave the way for corporations to make their maximum contribution to human welfare and to advance their own survival and prosperity[19]

One function of consultants is to facilitate the entry of inexperienced US or European firms into international competitive arenas. A related role consists in "opening doors" or "lubricating" the transition to investment and other operations in countries previously closed to Western capitalist firms. To illustrate, the legal firm headed by Samuel Pisar in Paris has, over a decade, brokered dozens of investment contracts between US firms and the Soviet Union.[20] Even experienced transnational firms feel the need to call upon consultants to survey ground unfamiliar to them. Not surprisingly, therefore, US corporations are now studying their possible entry into the People's Republic of China largely on the strength of early soundings taken by consultant firms.

International consultant firms are not limited to helping rich-country investors; they are also retained by firms, government agencies, or international organizations to perform studies in less-developed countries. In carrying out their daily chores they utilize for the most part "decisional" technologies: person-embodied expertise for diagnosing problems; abstract tools to simulate alternative policy courses and to weigh benefits and costs tied to each; and systematic disciplines the functions of which are to organize all available data on markets, employment pools, available technologies, sources of capital, or import legislation. Less-developed countries frequently retain the services of consultants to learn what technologies are available from the rich world, consultant firms serving as mediators between suppliers and utilizers of technology. For many firms lacking their own "technological gatekeepers," consultants serve as functional equivalents. But they are something more, veritable matchmakers, actively promoting the uses of new technologies by actual and potential clients who would not otherwise feel the need for same.

International consultancy is a highly competitive endeavor. Several top professionals have declared in interviews that the acute competition sometimes leads them to adopt, unintentionally, what they characterize as cynical and undesirable attitudes in the discharge of their functions. They usually assign blame to the system of "billability" under which they must work. The rule here is that professionals must maintain a certain percentage (usually running to about 70%–80%) of total work time which can be considered "billable" or chargeable to a specific client account. Consultants who do not maintain a high "billability rating" cannot qualify for promotion or be judged successful by their peers. They cannot afford the "luxury" of performing tasks which are not billable. Since competitive pressures lead individuals to struggle to sell their billable time, they rarely question the value assumptions underlying their own development models or those of clients.

Several Latin American government officials have declared in interviews that the main reason government agencies or private firms contract US consultants is not to obtain diagnostic or prescriptive expertise, which can often be found locally, but to take advantage of their easy access to consortium bank financing and their knowledge of how to prepare funding proposals for such bodies as the World Bank, the Organisation for Economic Co-operation and Development, and similar bodies. What is more, consultants enjoy the confidence of reputable corporations in the rich world. Thus firms and governments in the Third World view the endorsement of a prestigious consultant firm as a seal of approval for potential investment partners, financiers, or suppliers of technology. Among suppliers of technology, consultant firms which focus on diagnosis and systems design are the least vulnerable to restrictions or expropriation because their technology is intangible, embodied not in products or processes but in the accumulated "wisdom" of their personnel who enjoy contacts with a variety of clients, a broad gamut of governmental and international bureaucracies, universities, research institutes and foundations, labor unions, citizens' groups, and voluntary agencies of all types. Gatekeepers within large international consultant firms are in a privileged position to connect purely economic with purely technological with purely political factors bearing on investment and managerial decisions.

If the "sequence of dependency" outlined earlier accurately mirrors reality, good consultants may confidently expect greater institutional longevity in international arenas than may mere suppliers of capital, technology, or managerial skills. Their main expertise lies in domains of diagnosis and the coordination of all other factors with market dynamisms and their financial underpinnings. This explains why consultancy is at once so highly competitive and so highly remunerative. Not only do suppliers and utilizers of technology need (or think they need) the services of consultants but so also do most international funding agencies. The United Nations, the US Agency for International Development, the World Bank, and cognate institutions have institutionalized the use of consultants to study feasibility, sites, design, engineering, and financing. The World Bank goes even further; it makes wide use of consultants to help it process bids tendered by contractors in construction projects.[21]

The technologies wielded by consultant firms are not readily transferred; it is easier to train engineers to design and build dams than to teach feasibility experts how to evaluate myriad constraints on dam sites, design, and cost variables. Few Third World countries have fully understood that such "invisible" technological dependence may be a serious impediment to their efforts to minimize technological dependency. By and large it is consultants who set the broad frameworks within which most industrial and infrastructure technologies are transferred. Partly because fees paid to consultants often come from international funding agencies,[22] LDC governments have not closely analyzed their own degree of reliance on them. It can be stated without exaggeration that consultants are the cement or adhesive which holds the technology transfer nexus together: they serve simultaneously as mechanisms and as channels of transfer in high-priced competitive-market circuits. Even when international agencies subsidize the "transfer," the consultant is paid at professional fee levels.[23]

What emerges from habitual practices is a dismal conclusion, namely, that much technological problem-solving takes place without any veritable transfer of know-how. This is true not only in direct investment or controlled flows from parent firms to subsidiaries and affiliates but in many consultancy contracts as well. These contracts rarely include measures to assure that the pertinent expertise is communicated to the client.

The central lesson to be gleaned from this look at mechanisms and channels is that, although many governments treat the transfer of technology as an end in itself, companies engaged in transfer view it as a mere instrumentality serving their total marketing strategy. The preponderant role played by TNCs in technology transfers calls for further examination. Although they are seen by many as purveyors of technological salvation, so conflict-laden are their operations that judgments on them range from extreme condemnation to boundless approval. These operations, accordingly, are the subject matter of the following chapter.

4/Transnational Corporations and Technological Salvation

Recent studies on transnational corporations abound, and the purposes of this book are not served by reviewing them.[1] What does command our attention, however, is the oft-repeated claim that foreign firms benefit less-developed countries by bringing them modern technology. Yet, as one United Nations document states,

> the multinational enterprise is only one among several channels through which developing countries acquire proprietary technology from industrial countries. However, it is probably the most important and certainly the most controversial.[2]

Other UN publications meanwhile acknowledge the peculiar strengths of TNCs in domains of technology, marketing, finance, and management.[3]

The internationalization of production is a more significant effect of the expanded activities of TNCs, however, than is the size of their operations or the degree of control they exercise in host countries. Yet internationalization is inseparable from control. During an interview with *The New York Times*, José Bejarano, vice president in charge of Latin America at Xerox Corporation, declared that Third World nations need technology to abolish misery within their borders but lack the infrastructure to produce their own. Concluding that technological self-sufficiency lies beyond the reach of most less-developed nations, he advises leaders of these nations

> to recognize that solutions to technological problems are beyond their means. Officials of emerging countries can best foster technological development by enlisting the aid of the international business enterprise.[4]

Even developing-country leaders concerned over the dependency of their nations gain little from condemning Bejarano's stance or, as one militant party recently did, laying the blame for all the ills of technological culture on the Atlantic powers.[5] The harsh truth is that poor countries do need technology, and there exist few alternative sources outside the TNCs where they may obtain it. Business officials, conscious of the trump card they hold, can agree with Orville Freeman, president of Business International Corporation, that

> technology in the broadest sense—including material, managerial, marketing, organizational and other skills, as well as advanced technical information such as secret know-how—is at the heart of the difference between developed countries and developing countries.[6]

Moreover, as Raymond Vernon notes, even countries which are "developing" rapidly continue to require new technological inputs.[7]

TNCs will, in practice, continue to be major suppliers of industrial technology to a world in quest of "modernity" and its symbols. For this reason alone, the criteria used by TNC managers in exporting technology deserve examination by students of development. Without reviewing the general value systems of corporate firms, a topic surveyed in recent studies,[8] it is useful to explore how these general values are translated into specific criteria governing "technology transfers."

Suppliers of Technology: Their Criteria

Why do corporations based in rich countries sell technology to buyers in less-developed countries? One corporate official interviewed declared that "any US company will transfer technology if it gets paid for it and will keep control over the technology so that competitors can't get to it." His reply lays bare three criteria: monetary gain, the exercise of control, and competitive advantage over others. He argues that profit is justified because companies spend millions of dollars to develop new technologies. Research and development risks are very high; consequently, technology is not a free good but an expensive commodity.

Control by a specific corporation over the disposition of technology is closely tied to gaining advantages over competitors. Companies often implant technology in Third World sites to counter the potential moves of adversaries. The director of a US chemical producer declared it the policy of his company to "bring the latest technology wherever market possibilities exist." His firm built a subsidiary in Argentina during a boom in petrochemicals and plastics. The decision to locate in Argentina came "primarily from the threat of competition moving in from Germany and the United States." For several years production in Argentina proved unprofitable, although this trend was later reversed and plant expanded. Recently, however, certain product lines were dropped by the head office because "the company fell behind in technology. And you lose out if you can't stay with it in technology." Unless head offices control technological and market processes worldwide, they are not free to remove themselves from local technology arenas when these cease to be attractive.

Many international business executives, however, have a broader vision than that reflected above. The senior vice president of a prestigious engineering firm gave these reasons for seeking contracts in less-developed lands: (a) altruism: the desire to work professionally and for a profit in a way which can help the poor and save lives; (b) public relations: improving the corporate image by showing that the company is not a crass commercial entity interested only in profit and unconcerned with human suffering in the Third World; and (c) geographical spread: covering spatially differentiated markets when product diversification is ruled out because of specialization.

Most corporate personnel complain of being unfairly attacked by critics. They see their firms as *serious* companies which do not manufacture harmful products and which do provide excellent salaries and fringe benefits to their foreign employees and display greater ecological responsibility than their national counterparts. According to them, many Latin American governments favor a disastrous policy in the acquisition of technology. The cheapest way for LDCs to obtain technology, they claim, is to invite foreign investors to build full-ownership plants or take part in joint ventures. Third World efforts to purchase technology from noninvestors are bad, they add, because "the price is always too high."

A major struggle is brewing over the legitimacy of TNC roles in the Third World. Disputes center largely on issues of national control over corporate activities. TNCs frequently accept, however reluctantly, considerable interference with their normal practices of technology transfer. To illustrate, a Canadian mining company made an exception to its company-wide accounting procedures in recent years in order to counter restrictive legislation in Brazil, which treats technology payments as taxable profits and sets ceilings on royalty payments which fall below the company's acceptable minimum.[9] Accordingly, the company's Brazilian operations are simply not billed, as are other subsidiaries and affiliates, for R&D expenses incurred in the home country or for technological services provided in Brazil. These costs are "picked up" in some third country or carried as a debit—*sine die*—on the books of the Brazilian units. This practice illustrates how the operative criteria adopted by TNCs in selling technology to LDCs relate to overall firm profitability. Yet great flexibility is the rule within "enlightened" corporations, especially those which have survived many generations of troubled times. According to them, most restrictive national legislation can either be circumvented or "ridden out." Their thinking suggests still another, albeit implicit, criterion invoked by suppliers of international technology: the desire to slow down the process of coalition bargaining carried out by Third World countries.

The director of one large international marketing firm justified his own opposition to such coalitions in these terms:

> The persistence shown by many in wanting to treat Latin America as a continental entity confirms the existence of an international bureaucracy, mostly of leftist origins, who are the chief agents (unconscious) of precisely what is anachronistic in both capitalism and socialism.[10]

Business consultants frequently warn managers in client firms of a possible trend in the direction of such coalition bargaining and suggest how Third World concerted actions can be circumvented.

More importantly, however, corporate decision-making in the home country decisively affects local technology arrangements in the host country. One spokesman for a large transnational firm explained that his firm's decision to invest in Latin American Country A was dictated by the desire to maintain "objectivity." According to him, most foreign companies feel attracted to invest in neighboring Country B. "But my company can't risk being viewed as tied to one country (B), or to one set of interests." Because the company believed that its interests transcend national biases and that its regional offices should not be linked to the destinies of any single country, the company installed its regional office in Country A, which is geographically close to Country B, where it has more numerous and larger plants. Besides, "it is not good business to be too close to local managers." This decision created problems at the technical level. It would have been more "efficient" for the company to locate in Country B, but the company decided to subordinate efficiency to the criterion of managerial "objectivity."

Barnet and Müller consider that "no aspect of the technological superiority of the developed world is more important than its mastery of the techniques of ideological marketing."[11] But ideological marketing need not be conducted in manipulative fashion; nor must corporate managers deliberately mystify foreign clients. It suffices that these managers become "true believers" in the beneficence of their overseas operations. Because many of them *are* so convinced, they are troubled by what they judge to be politically inspired "distortions" of a purely business issue.

Indirect confirmation of this view was obtained during an interview conducted in Santiago on 13 April 1973 with a top economic planner under Salvador Allende in Chile. This official complained that Chile "was negotiating with less strength than before. One reason is that whole packages are negotiated; and, therefore, in order to get the package, Chileans make concessions in matters pertaining to technology." Three days later, a Chilean economist who had worked with the predecessor Frei regime refuted this impression by insisting that "technology is technology" and that it matters little who supplies it. According to him, "Chile now imports technology from socialist countries under more rigid conditions than former contracts with the United States or Europe. The USSR 'donates' turn-key plants [for example, a fish-processing plant, or a factory to make prefabricated housing], but then follows this up with tied credits and with tied technical services." Such procedures, in his view, were not conducive to Chile's gaining technological independence.

These and other observations from government sources during a period of socialist experimentation cast light on other important aspects of international technology transfers. Thus during the final months of the Allende government, one UN administrator living in Chile voiced the opinion that "Latin American countries may have only a few more years in which to devise alternatives to merely integrating into the world capitalist system of trade and technology exchange. If they fail, they will isolate themselves from modern technology and the pool of trained people." The knowledge that their prospective clients fear just this possibility is never absent from the minds of corporate managers as they bargain with host governments eager to minimize national dependency. Whatever the wishes of the latter, there are limits to their possibilities of autonomy. Indeed many transnational managers, including those working within Andean Pact countries, exhibit impatient disdain for those who legislate restrictions on TNCs within their borders. They tend to dismiss them as "intellectuals who have no experience, and who have now become bureaucrats," revealing one latent attitude prevalent among business decision-makers: that they need not take "seriously" academic or governmental students of technology. Such decision-makers, however enlightened, have no respect for those whom they brand as victims of their own unrealistic generalizations or ideological illusions. The universe familiar to these decision-makers is that of competition for money and for bargaining power; hence, they profess to respect only those who display knowledge and experience in these arenas.

No probe of corporate criteria in technology transfers is complete unless it evokes issues of geographical concentration. Large firms, it is true, seek geographical diversification in investments and licensing outside the United States, but they also prefer to concentrate their efforts, at least initially, on one country or region. Firms working in engineering, construction, and social infrastructure (sanitation, water supply, health units, etcetera) are attracted to dollar-rich Arab countries for obvious reasons. First, these Arab clients have abundant foreign exchange and need not depend on international agencies for hard-currency funding. (Indeed the companies just mentioned find international funders to be insufferably slow and cumbersome; moreover, the competitive bidding procedures they impose on Third World clients interfere with the "special arrangements" these companies are fond of making with client states which "trust" them.) A second reason is that dollar-rich countries do not insist on certain "uneconomical" clauses imposed on consultants by world funders. To illustrate, one engineering official complained that the World

Health Organization occasionally requests the services of a single expert for a prefeasibility study but sets a ceiling of $5,000 for his services. This, said the engineer, is too low to "get a good man." He further complained that world bureaucracies have generally unrealistic fee schedules. As a result, they "end up getting second-rate Central American [*sic*] engineers." To counter these nuisances, he explains, his company and most competitors have established a dollar limit below which they will not accept a job. A third reason why consultants of the type described favor concentration in one country or in a cluster of nations in the same area is that their high-quality people can be put to double or triple project-use, all the while minimizing expenditure. And it suits them better to have contracts directly with governments which can pay rather than to rely on world funding agencies which pursue a policy of "spreading their loans around geographically."

Engineering and design firms have explicit preferences in clients. They view with disfavor the plea made by many LDC governments that they engage in substantial training activities of counterpart personnel in the course of "getting the job done." This imposition, they argue, reduces efficiency and brings engineering costs up. Moreover, it makes the intended beneficiaries of a project—the poor of an Indian city or the peasants in the Iranian countryside—wait longer for the delivery of the services (water supply, electricity, hospitals) than would otherwise be the case. One engineer puts it this way:

> To provide technological transfer is a hindrance to the achievement of the immediate objective, shared by our company and our client government. To insist on a transfer via training inhibits the company's performance of its given assignment.

Another criterion which engineering, design, feasibility, and consultant firms apply in transfer decisions bears on costs. "Quality" suppliers of expertise encourage LDC clients to hire experts on the basis of competence first; costs, they say, should be discussed only after the choice of firm is made. Their assumption is that to look at costs *first* places competent serious firms at a disadvantage in the bidding and the preliminary review of candidates. The general presumption is that less-skilled competitors offer "bargain basement" services to unsuspecting clients by sacrificing quality. Psychological "selling" of quality know-how is as central to the strategy of international suppliers of technology as it is to those who sell products or processes. So true is this that some consultant and engineering firms try to score an additional "selling point" for their higher fee schedules in one of two ways. One approach is to present elaborate calculations in which "quality" is weighted in total costs. The result is that in terms of "pound for pound" of quality, as they put it, they end up being *cheaper* than alternative firms. They often aver, too, that they do not calculate professional service charges by hourly rates or according to time spent by diverse categories of professionals employed on a project. The reasoning put forth is that such accounting diverts attention from the main objective, which is to finish the report, to render the design operational, or to map out a precise investment strategy; tangible results like these are adduced as the best measure of money well spent. Hence the preference of high-level consultants for working within stated budgets rather than on a cost-plus-fixed-fee basis. A second approach is for a firm to admit that it is indeed more expensive than others, but that it offers more solid guarantees and more reliable and experienced personnel, so that, in the long run, "to cut corners" on quality is a diseconomy because the job will be poorly done or will have to be "patched up" soon after completion. So confident are "quality" firms of the persuasiveness of their criteria that they sometimes "risk" losing a lucrative contract by being "too honest" with LDC clients in preliminary negotiations. They tell them outright that the proposal is wasteful, inefficient, or doomed to failure. Even when they lose the contract, however, they usually affirm that they would be equally "honest" the next time, so confident are they of being proved right by experience. To quote one engineering consultant, this is the reason that "we prefer to work with professionals rather than with amateurs: a shared universe of professional values and criteria can be presumed to exist."

One form of technology transfer which has not been widely exported from the United States to less-developed countries is that wherein a technology innovated in one field is adapted to another.[12] Firms like ABT Associates play an important "matchmaking" role in searching systematically for industrial and commercial applications of technologies initially developed by the National Aeronautics and Space Administration (NASA) in its space programs. Mediation is performed also by universities, usually with funds obtained from the federal government. The imagery which "matchmakers" use to describe their function illustrates an important aspect of technology transfers. They justify their role by claiming that those who are knowledgeable about markets in one field are usually not technologically knowledgeable in some other field; and the converse, they claim, is likewise true: people with technological knowledge in one field are rarely aware of market opportunities for that technology in fields other than their own. Hence the need for a "detached third party who can help potential creators and final consumers of technology talk to each other without letting their specialized blinders or vested interests stand in the way."

One view widely held in corporate circles postulates a tight correlation between the stability and size of the market and the ease of communicability of technology. That is, if the market is relatively stationary, transferability of technology is high and goes smoothly because most firms possess the same level of know-how and conditions throughout the market are similar. The opposite situation prevails where market growth is rapid and high, a circumstance which creates much latitude for new customers and products. In this instance, the likelihood that any two producers will operate at identical levels of technology is slight. Consequently, one firm is more apt to experience more difficulty in transfers than its competitor. The conclusion to be drawn is that the greatest competitive advantage derived from technology occurs in growth situations.

The theory under review also examines questions of scale and size. Studies conducted by the Boston Consulting Group, a prestigious international consultant firm, suggest that, within broad limits, whenever one doubles the number of production units, unit costs can be reduced by 10%.[13] Armed with this knowledge, many firms propel their search for new technologies in order to gain a more favorable market position. They try to force the market to expand, and in order to do so, they will accelerate R&D investment. Implicit here is a dynamism parallel to that found in Vernon's "product cycle" theory. Indeed the criteria of technology suppliers are best understood by recalling that they treat technology not simply as an aid to production but also as a product in its own right. Thus pharmaceutical firms sell technology in older products to Third World countries in order to gain market entry for their newer products. Transfers of this type are not risky; on the contrary, they are viewed primarily as incidental ways of deriving supplementary income from know-how which may well be on the way to becoming nonproprietary knowledge.

I once asked a management strategist, "What is the most proprietary kind of knowledge in existence?" His intriguing reply was, "The most proprietary knowledge is the kind nobody knows anything about. For example, there is a sensitive process in one metallurgical plant where the trick is to turn on the lights when it's exactly 300 degrees. Only three guys [*sic*] in the company know this trick. And, obviously, this is nonpatentable technological know-how." Examples like this confirm the idea advanced by Louis Wells that "there is no dividing line between proprietary and nonproprietary knowledge."[14] Smaller firms patent and license their technologies more readily than larger firms because there is little else they can do with their technology: they lack the capital and personnel to apply it directly to increased production, a far more solid font of additional profit.

Transnational corporations do share a basic set of criteria in technology transfers. Technology will be transferred when it leads to heightened market penetration or control. Individual companies doubtless seek profits on their transferred technology, but more important to them than immediate profits is the conquest of a favorable strategic position whence to control future shifts in their overall marketing position. Speaking practically, this means that if technology must be licensed if that is necessary to enter a market, to keep competitors out, to diversify product outreach in existing markets, or to gain a more intimate knowledge of local market adaptations required, then it will be licensed. General agreement exists among corporate decision-makers that income derived from technology transfers is a relatively minor source of total revenue. Yet absolute prices are high; technology is expensive and risky to develop. Therefore, it is expensive.

Why do overall market strategies dominate the thinking of TNC officials? As one manager explained, "Unlike public organizations, the corporation has an accounting system that forces it to interact with its markets every day. This is its greatest strength." This strength is now being challenged in domains of technology transfer by governments, international agencies, political groups gaining support of international public opinion, unions, and consumers themselves. All are "chipping away at the freedom of maneuver enjoyed by companies." This, say TNC officials, is bad because it leads to lower efficiency and, ultimately, to reduced economic growth. Conflicts inevitably arise: TNCs as suppliers of technology insist that they need maneuvering room if they are to provide LDCs with what is asked of them—the dynamism of growth. So goes the argument.

The claim made, implicitly at least, by transnational suppliers of technology that they are vectors of technological salvation cannot be judged, however, unless their criteria as suppliers are contrasted with those invoked by Third World purchasers of technology.

Criteria of Purchasers of Technology

Buyers of technology in international exchanges have diverse objectives, owing initially to the fact that these buyers themselves are a heterogeneous lot, ranging from affiliates or subsidiaries of transnational corporations to state-owned firms, private national firms, and third-country firms. It is useful to speak first of the objectives of poor-country firms in acquiring foreign technology. Only afterwards shall I list the benefits host governments seek in their technology import policies.

Subsidiaries and affiliates do not generally place a high priority on cost-minimization in their acquisition of technologies. Most of their purchases are made from parent firms—precisely the kind of transactions in which abusive transfer-pricing and over-invoicing occur most frequently.[15] These exchanges are usually conducted as package deals in which technology is tied to restrictive marketing clauses, to royalty payments for trademarks or goodwill, and to contractual obligations to purchase raw materials or so-called "intermediates" from the parent firm, often at guaranteed prices. These practices frequently prevail in contracts between TNCs and client firms; but in parent/affiliate relations they come close to being the norm. Moreover, local managers of affiliates or subsidiaries are judged in light of overall performance standards set by the parent company. Quite predictably, therefore, it is to the parent firm that their professional loyalties are ultimately given.[16] By and large, these managers hold the same value system as that favored by suppliers of technology; the primary concern of both is to assure a regular supply of technology that "fits" the production requirements of local firms. And by definition, such technology is the kind developed and sold by the parent organization. Although full assimilation of that technology within the local firm is important, it matters little whether such assimilation takes place in the person of local or foreign personnel. What is crucial is that production and marketing harmonize with overall company strategy. Local managers sometimes wield enough influence to bend parent decision-makers to local needs or to special technological adaptations. Even these managers, however, are rarely moved by values other than maximizing their own efficiency, competitive advantage, or career aspirations.

On the other hand, managers of locally owned enterprises, private and public, operate from different motives. The aura of prestige surrounding "modern" technology weighs heavily in their decisions over desirable sources and kinds of technology. Quite often they sympathize theoretically with the goals proclaimed by their governments: lowering costs, optimizing local materials, and reducing outside dependency. Because these local enterprises need to be competitive, they look in practical terms for technologies for which are assured standard quality, reliable delivery, guaranteed maintenance and supervisory services, and favorable financing terms. In order to obtain these benefits they agree to pay high prices, to remain dependent on outside suppliers, and to sign package deals even in the face of government policies expressive of opposite priorities. A government planner in one Latin American country explained that even state-owned firms often sacrifice their independence if technologies available from TNCs are of high quality, especially if hard-currency financing accompanies their purchase. In his opinion the main incentive for any local entrepreneur to *avoid* importing foreign technology is the shortage of hard currency. Suppliers of know-how, alert to this constraint, frequently provide favorable financing. The specific criteria applied by local decision-makers vary widely as they contemplate acquiring product, process, or design technologies. Local firms have a strong incentive to develop local product technologies which bring their goods into closer line with local demand, tastes, and preferences. But in the case of process and design technologies, there is less pressure on locals to avoid standardized outside imports; other considerations—reliability, prestige, assured delivery—weigh more heavily.

The interests of LDC governments and firm managers coincide in one important realm, however: their own personnel to master acquired technologies.[17] Local firms do not want their in-house technology to remain too dependent on foreign experts. Hence they place a high premium in technological contracts on training their personnel, often requiring that employees be trained in the factory of the supplier and that foreign experts periodically visit and instruct their local technicians. To obtain this service, however, they make concessions to suppliers in transfer-pricing, package deals, and marketing restrictions.

Plant directors in LDCs identify the pressures which perpetuate their dependency on outside suppliers. First, it is very disruptive for them to break with patterns once established: sales go down, productivity drops, skilled personnel leave, and their market position is weakened. Second, they are usually dissatisfied with available alternatives to technologies supplied by world-known corporations: quality is inferior, delivery dates are uncertain, quantities are insufficient, and political or personal considerations interfere with a purely "businesslike" relationship with local suppliers. A third powerful constraint is the shortage of trained negotiators. Negotiating good contracts poses an array of complex problems requiring legal, technical, economic, and conjunctural skills. Most small and medium-sized firms already rely on the services of patent attorneys or agents to negotiate patents, royalties, trademarks, and licenses. The managers of these firms fear that payments to such middlemen would increase drastically if they prematurely severed their bonds with "familiar" outside suppliers of technology.

A general lesson gleaned from examining constraints facing purchasers of technology is that incentives play a decisive role. LDC governments can achieve little success in reshaping patterns of technology acquisition unless they offer new incentives to local users of technology. Incentives are discussed at length in Chapter Eight, which deals with negotiation strategies. But a rapid look at the criteria invoked by the governments of countries importing technology is now in order.

The first objective of LDC governments is to harmonize acquired technology with the objectives of national development. Although few governments are fully coherent in pursuit of this objective, it nonetheless stands as the professed goal of all, for "having" technology is seen as part of being modern or developed (however uncritical may be the assumption that imported technology furthers the cause of national development). Other criteria emerge from the perceived need of governments to control the "strategic" sectors (or commanding heights) of their national economies. Two special considerations here enter into play, namely, the governments' felt needs to assure (a) national control over industries related to defense or military production, and (b) autonomy in sectors considered vital to the economic welfare of the country. While defense industries are nearly universally and certainly traditionally considered "strategic" by all nations, countries which depend heavily on the export of a few commodities for foreign exchange will also include those sectors (for example, cocoa, jute, timber, cotton, palm oil, coffee, bauxite, copper, or tin) in such a definition. The choice of an industrial strategy may also lead to the classification of other sectors as strategic, as seems to be the case for electronics in Taiwan, petrochemicals in Algeria, and textiles in India. Stretching the definition yet further, many LDC governments also count as strategic (for noneconomic reasons) basic social sectors: health, housing, education, communications, banking, and transportation. Unless relative self-sufficiency is gained in these domains, they argue, the country will be too vulnerable to decisions made by outsiders; moreover, these social goods bear directly on employment and the basic survival of the masses.

Still other criteria are equally important to technology-importing countries: lower acquisition costs, optimum use of local resources, and reduced dependency on the outside.[18] The last criterion is paramount in the sense that national leaders ultimately seek the capacity to reach autonomous decisions on technology. But such autonomy is impossible to achieve unless the second criterion is met, that is, unless local finances, materials, and human skills can be put to optimal use. As to the first measure—lower acquisition costs—such can be negotiated only through increased bargaining strength. Dependence does not mean the reliance on outside suppliers of technology, but rather a country's inability to control the direction, speed, and social effects of technological evolution. Dependency's opposite, autonomy, implies the ability to shape a technology policy which serves the national development strategy and central value options of a society. Planners who enter the technological arena have two essential objectives. One is to acculturate the masses to an increasing receptivity to, and familiarity with, technology. For unless technology is widely assimilated by a population, social stratification will become, or remain, hierarchical and elitist: engineers and technicians will monopolize all decisions affecting production. Especially where such goals as greater social equality and decisional participation are important to development strategy, serious efforts must be made to disseminate technology. To this end, the focal points through which technology enters industrial firms need to be organized, along with suitable education and occupational incentives. The second objective of planners is to choose appropriately from a wide range of technologies. The notion of "appropriate" choices of technology, in turn and likewise, implies a pluralistic approach to planning. That is, it becomes essential to identify sectors and activities where so-called "appropriate" or intermediate technologies best ally the need for higher productivity with equally important needs to provide employment, to utilize local materials, to associate a poor populace with its own development, to conserve depletable natural resources, and to reduce dependence on foreign currency.[19] Clearly, then, according to such measures, the option lies not in a choice between allegedly "soft," "appropriate" technologies versus inappropriate, "high" technology. Rather, the wise course is to employ the entire gamut of instruments ranging from improved traditional technologies to others which are modern but small-scale, labor-creative, and indigenously developed; to others which are second-generation or "obsolete" technologies imported from developed countries;[20] to still others which are the most modern of all; and perhaps even to others which do not exist anywhere yet but which must be developed *ex ovo* to suit special needs.[21]

Much light is thrown on appropriate technology choices by the practices adopted in the People's Republic of China. According to recent studies the Chinese have adopted capital-intensive modern technologies in capital-goods industries so as to become internationally competitive and acquainted with modern innovations.[22] But in industries which manufacture consumer goods (processed food, textiles, bicycles, sewing machines, transistor radios, etcetera) other priorities take precedence: maximizing employment and involving workers in production and research decisions. Some degree of efficiency is sacrificed to the preference for labor-intensive technologies which utilize local parts, know-how, and currency, with little regard for internationally set quality standards. An "appropriate" technology policy thus covers the entire range of tools, organizational systems, and work modes, with the degree of "appropriateness" judged sector by sector, industry by industry, and product by product in accord with broad value options taken by society and expressed in precise development strategies. A creative mix of diverse approaches is deemed indispensable, along with great flexibility to cope with new constraints—hence the high reliance placed by the Chinese on the process of trial and error and their rapid correction of unfavorable trends once detected.

Many government planners fail to utilize the diagnostic tools they need to make "appropriate" choices. One reason for this, already explored, is their lack of ideological clarity as to broad social goals. But another is the absence of an integrative principle for screening and evaluating technological options. One such principle which may prove useful is the "Sabato triangle," a model for technology policy formulated by the Argentine physicist and metallurgist Jorge Sabato and his coworkers.[23] The model aims at creating practical linkages among research, production, and development-policy actors. Its underlying image is the triangle, a geometric figure with three interconnected vertices. One vertex represents governmental decision-makers; the second, producers; the third, scientific and technical researchers. Each vertex must be linked by a flow of information with the other two; each must also take initiatives in demanding or supplying technology. Factories must have access to, and influence on, laboratory or university researchers. Conversely, governmental planners must be able to influence which technologies manufacturers will use. Unless circulatory flows link all elements of the triangle, there can be no sound incorporation of technology and science to national development. Sabato offers two interesting historical examples which helped him formulate his imagery.[24] The first is the invention of the stirrup in the early Middle Ages, a breakthrough which instantly transformed horses into fantastic weapons of war. Suddenly rulers needed more horses and more land on which to raise horses. Given existing land-tenure systems, this meant the need to expropriate more land—from churches and feudal lords. What seemed to be a mere technological invention thus proved to be a potent agent of historical change. Pre-World War II Germany provides Sabato with his second example. Since Germany was rich in zinc but poor in copper, Hitler ordered industrial researchers to find ways of making automobile carburetors out of zinc instead of copper. Thus stimulated, research and productive actors in the "triangle" proceeded to invent a zinc carburetor.

These cases illustrate the Sabato principle: if an institutional triangle functions vigorously, technology will contribute directly to development. But if the triangle is absent or weak—that is, if the apexes are not connected or if infrastructures are deficient—any technology locally produced or imported will make but a slight contribution to development. The policy conclusion Sabato reaches is that triangles must be deliberately set up for each major sector within domestic economies as well as for absorbing and disseminating technology acquired from without. In most cases this prescription requires new government initiatives to endow the research infrastructure with more resources. Or it implies the more arduous task of redirecting present research infrastructures away from pure research unrelated to productive needs toward organic linkage with productive requirements. The concept of the "Sabato triangle" enjoys widespread acceptance among Latin American students of technology—Sagasti, Vidal, Vaitsos, Kamenetsky, Herrera, Giral, Kaplan, Wionczek, et al. Furthermore, organizations such as the Andean Pact, the Organization of American States, and several national governments take it as their reference point in formulating national technology policies. The model's strength lies in its great simplicity and obvious practical applicability. The "Sabato triangle" is thus highlighted here because it serves increasingly as a basic tool utilized by Third World governments whose firms import technology from transnational corporations. The model also enables policy-makers to identify and manage conflicts of interest between suppliers and purchasers of technology.

Conflicts of Interest

Friction between transnational corporations and importers of technology centers on two basic divergences. The first is traceable to initial perceptions. TNCs view technology as an expensive commodity with a very short commercial life. Consequently, it must be sold at a high price. Many persons in LDCs, however, contend that research and development conducted by TNCs is amortized in their home-country markets. Therefore, they claim, technology exported to them ought to be modestly priced. Some in the Third World further argue that technology should be transferred automatically along with any direct investment made by foreigners. According to them, LDC governments are justified in placing ceilings on license payments for technology. Corporations, on the contrary, regard such ceilings as arbitrary interferences with normal market mechanisms.

A second zone of conflict focuses on control over imported technology. Transnational firms resent efforts of host-country governments to wrest control away from them. They regard such measures as required registration of all licenses, legislation on royalty arrangements, and payment ceilings in technological contracts as hostile gestures. At the very least, TNCs resent the extra expenses they must incur to meet paperwork requirements; at worst, they view restrictions as discriminatory and unjustified limitations on remittances. In truth, license registers lack any power unless they are backed up by sanctions vested in central banks or other monetary institutions having the authority to freeze transfer of payments. But corporate officials complain that administrators of technology registers make discriminatory judgments. Indeed registry officials do decide whether comparable technologies are available within the host country; if they are, foreign suppliers can be declared ineligible. Yet the same officials can make exceptions on various grounds, by determining, for example, that local teams are not available immediately, that they cannot provide financing as part of the "package," that the registry lacks the personnel to supervise quality, and so on. Champions of registers nonetheless defend their necessity, alleging that without them vulnerable governments cannot protect themselves against exploitative prices, excess packaging, and high payments for intangibles like trademarks or goodwill, all of which they brand as "fictitious" technology. Disagreements at the theoretical level reveal little, however, about daily practice. In fact, large TNCs, in order to protect their reputation as good corporate citizens, usually comply dutifully with local registry requirements, but at the same time, they continue to seek ways (usually legal) to obtain the equivalent of their normal payments from clients. And company spokesmen unhesitatingly declare that the LDC governments will choke off their supply of needed technology "if they go too far in placing restrictions."

There is no need to repeat what others have written on transfer-pricing or on multiple bookkeeping, another conspicuous source of friction.[25] United Nations studies on TNCs have now brought these practices out in the open, and the various codes of conduct now beginning to circulate evidence the general concern over this issue.[26] Gaining strength is a movement to outlaw transfer prices which are higher than "arms' length prices" charged to third parties. Although TNC officers assert that transfer-pricing is justified in terms of the overall exchanges of which technology is but a part, this practice will not escape the scrutiny of LDC governments once they fully understand how internal pricing works in practice.

More basic than specific conflict is a general disagreement over the role of transnational corporations in channeling technology transfers. Corporations employ a rhetoric which portrays them as purveyors of technological salvation. The "line" is that if modern technology is adopted, misery in the Third World will be abolished, productivity will increase, and everyone will be better off. And further, TNCs are the best channels for bringing technology to poor countries because of their global organizational skills, their ability to mobilize resources quickly, their skills in recruiting personnel from all cultures, their capacity to respond quickly to opportunities, and their massive investments in R&D without which new technologies could not be generated. While agreeing that TNCs no doubt possess these advantages, one can still legitimately doubt whether the technologies they supply are well-suited to abolishing the poverty of masses in poor countries. A later chapter analyzes in detail the high price exacted by technology transfers. Worth noting here is the hypothesis formulated by the Brazilian economist Celso Furtado, namely, that technology transfers conducted by TNCs increase social inequalities among classes.[27] The reason is simply that corporate managers calculate efficiency in a way which favors products and services designed to meet the purchasing power of privileged classes in poorer countries. Their technologies are not "efficient" in terms of satisfying the needs of the poor.

Many in developing countries doubt that technological salvation comes via transnational corporations. In agriculture, to cite one case, modern technology tends to reinforce the gains reaped by large agribusiness firms at the expense of peasants or small farmers.[28] Criticism of TNCs as bearers of technological salvation leads to several different challenges to them and prescriptions for control. Some assert the right of less-developed countries to displace, at least within their own territories, TNCs as the main controllers of vital technology. Partisans of "soft" technology challenge assumptions held by corporate technologists as to optimal scale, criteria for internalizing factors in the efficiency calculus, and a preference for centralization. Instead they seek smaller technologies which are supportive of local values, protective of nature and natural resources, less dependent on expensive raw materials, better adapted to handling by unskilled people, and highly labor-intensive; it is from these "alternative" technologies that they await true development. Others, however, arguing that transnational corporations must continue to be the main providers of technology because no effective alternative can be found, claim that TNCs can and should be regulated to respond to the values defended by "soft" technology. These critics also urge Third World governments to assimilate imported technologies more critically and to diffuse them beyond the confines of the commercial partner of the exporting TNC firm.

A word must be said regarding the values of international business personnel.[29] One image is central in explaining the infatuation of corporate managers with dynamic change: the notion of *challenge*. In most interviews with corporate officials, the term *challenge* appeared in their answers to the question: "Why do you sell technology to the Third World, thereby entering an arena you characterize as difficult, uncertain, and at times even dangerous?" If one term comes close to signifying universal praise for a job, it is that the job is *challenging*. Engineers say they are attracted to difficult, although potentially frustrating, research tasks because these pose a challenge which transcends routine. Managers and consultants risk ulcers and nervous breakdowns because their jobs surface an endless series of challenges which test whether they "measure up to their potential."

Challenge, then, is the value repeatedly invoked to justify corporate involvement in uncongenial Third World sites. Yet challenge-seekers take for granted other rewards: high salaries, recognition from peers, company promotions, and tangible fringe benefits. Therefore, challenge must be seen as a symbolic value to which one appeals in order to legitimate placing one's talents at the service of mere profit. Challenge-seeking is the *moral category* which provides incentives to action, thereby bridging the distance between mere professional efficiency and a desire to "help mankind." Other theoretical categories are doubtless also at hand: the notion that economic growth, greater efficiency, and modern technology all greatly contribute to human development. But one cannot get emotionally aroused over maximizing efficiency unless the effort is a challenge, an exhilarating game. No matter how routine or trivial an organizational task may be, it can be endowed with a value rewarded by the corporate system. All that is needed is to subsume it under the heading of "challenge."

Challenge stimulates corporate energies, reinforcing the commitment to growth, expansion, and unceasing technological innovation. To develop or sell new products and new processes can be seen as a challenge, especially in poor countries where markets are small, skilled personnel are scarce, legislation is restrictive, and logistical problems are the daily norm. Two qualities serve as sources of challenge to business: difficulty and the uncertainty of "winning the prize." Judging from laments of transnational managers that "uncertain" conditions in less-developed lands are the greatest obstacle to smooth functioning, one must conclude that difficulty is the preferred font of challenge. Here an interesting dichotomy arises in the minds of corporate personnel accustomed to operating in developed-country home bases. When challenge springs from a greater difficulty of achieving success—and this is the ordinary situation in underdeveloped lands—even successful handling of the challenge can be less rewarding, materially, than elsewhere. Where challenge comes not from difficulty but from uncertainty there is a higher risk content to decision-making. The element of play acquires salience: games take on a special appeal to powerful decision-makers. Challenge cannot thrive in the absence of power, real or desired, for it is in "power games" that victories are most highly rewarding and defeats most stigmatizing. Nevertheless, the second kind of challenge is diffused widely throughout many institutions only after a certain level of wealth is attained: it is, in a word, a luxury. In terms of creativity, however, challenges founded on uncertainty are more conducive to technological breakthroughs than those founded on mere difficulty.

The workings of "challenge" as a kind of hidden meta-criterion underlying corporate *drive* suggest an analogy with William James's "moral equivalent of warfare." Corporate managers, planners, and technicians are socialized into perceiving challenges wherever there are new products to be made, new profits to be gained, new markets to be conquered. Just as in the past warfare stirred men to display assertive qualities of audacity, physical courage, and triumph over the fear of death (traits which James wants to encourage through means other than war), so too does the corporate system socialize its managerial and technological soldiers around challenges attaching to competition; conquest; and the aggressive development of new products packages, and selling messages. Of course a functional equivalent of merchandising warfare could locate challenges in different values; it could conceivably enlist energies and talents in the quest for cheaper, more durable, and less wasteful ways of producing goods needed by the masses.

Positing this option leads us back again to a consideration of the vital nexus binding basic value options to development strategies to specific technology policies. In order to understand conflicts between suppliers and buyers of technology, one must work backwards and unravel the skein of this nexus. Thus, if one dislikes a policy, one must also question the undergirding strategy and ultimately reconstruct the values. Hence if the policy favored by transnational corporations in technology transfers is to be reversed, their overall marketing strategies must likewise be altered. More fundamentally, corporate personnel need to develop allegiances to new values. The *locus* of challenge, in short, must shift from infatuation with material growth, quantity, and merchandising manipulation to the ambition of assuring the achievement of integral growth. Integral growth does not place material triumph over personal communion or social justice. It is concerned with the quality and durability of materials and technologies, with meeting genuine needs of all humans rather than with flattering the wants of those who are "natural" customers of corporations because they have purchasing power.

* * *

To summarize, suppliers and purchasers of technology obey different criteria. Suppliers want technology transfers to be *lucrative, un*fettered by *extraneous* (that is, by nontechnical) considerations, and *congenial to their habitual modes of operation.* Indeed they wish to use transfers as means to gain themselves footholds in diverse markets; to initiate offensive and defensive measures against competitors; to gather additional gains from research and development already conducted or in process; to counter domestic pressure in home countries over ecological or labor conditions which presuppose ready absorptive capacity of technology at the receiving end; and to take advantage of international financing of contracts.

A contrasting list of attractions moves purchasers of technology. Importers want know-how which will help them remain competitive in local and/or international markets; solve their problems better; buttress their image of "modernity"; gain entry into the developed world's pool of managerial, financial, and technical expertise; contribute to their wish to industrialize; produce and merchandise new products or services; make greater gains; gain professional mobility as an "international" technician; and protect vital links to the outside "developed" world.

On balance, are TNCs bearers of technological salvation? A review of conflicting criteria adopted by TNCs and LDCs for transferring technology reveals that neither technology nor salvation comes very easily. One glimpses the high price paid by underdeveloped countries for imported technology. Before inquiring systematically into this price, however, several case studies drawn from my field research will be presented. These illustrations add concreteness to the abstract issues expounded thus far.

5/Case Studies in Technology Transfer

The present chapter illustrates, with case studies drawn from my recent field research, how and why priorities of suppliers and utilizers of technology diverge. At times, differences cannot be reconciled. Even when full harmony is unattainable, however, valuable lessons can be learned regarding transfer negotiations. One key to success is engaging, in early stages of negotiations, in critical discussion of the value assumptions of partners to transfer contracts. Although debate at this level is full of friction, it reduces misunderstanding at later stages.

The cases described here are neither necessarily typical nor representative of any statistical class of phenomena. They do, nonetheless, illustrate the dynamics of international technology transfers and negotiation strategies. In most instances here presented, all partners to the transfers were reasonably satisfied, but none of the cases is an "unqualified success story" which can serve as a paradigm for other efforts. What emerges more clearly from these studies is that, even in achieving a relative "success," certain values must be sacrificed.

Various institutional actors are included in the cases chosen: universities, government agencies, consultant firms, manufacturing firms, and peasant villagers in a mountainous country. The roster of cases includes a university project for water-basin development in Argentina, a consultant study on cold-food systems in Brazil, licensing arrangements in an Argentine shipyard, overall operations of a US precision-instrument firm in Latin America, general remarks on value conflicts in tourism, and miscellaneous short cases.

These exhibits reveal how technology is both a destroyer and a promoter of values and an instrument for creating new bonds of dependency even as it removes old constraints. The link between technology transfers and market competition is likewise brought to light. Finally, the cases show concretely how transfer mechanisms operate and what roles transnational corporations play in moving technology from one society to another.[1]

Case 1: Water-Basin Development in Argentina

The Massachusetts Institute of Technology (MIT) has conducted a "technology transfer" to the Sub-Secretariat of Water Resources, an agency of the Argentine government, with a view to achieving three objectives:

- to construct a framework of comprehensive planning suitable for use in future water-basin development in Argentina and elsewhere
- to train a group of Argentine professionals in the theory and practice of multipurpose water-basin planning
- to prepare an integral development for the river Rio Colorado using these methods

The original two-year contract expired on 30 September 1972 but was renewed for two more years. There is no need here to relate contractual details or specifics of the MIT action plan.[2] What is important, however, is to review briefly the rationale for what MIT Professor David Major has termed "a successful transfer of systems technology from one country to another."[3]

One important element consisted of conducting "trial runs" of multi-objective or multifunctional water-resource planning. Investment *criteria* were drawn up *to optimize* a combination of objectives—net contribution to national and regional incomes and harmonization of social, environmental, defense, and economic goals—sought in the specific programs. MIT designed its approach to be even broader than so-called multipurpose planning in water-resource management, a term which evokes multiple benefits expected from such projects—irrigation, hydroelectric power, and water control. The multifaceted approach was thought vital to the Rio Colorado basin selected by Argentines in joint negotiations with MIT in part because the river flows through five provinces with different needs: Mendoza, Rio Negro, Neuquên, La Pampa, and Buenos Aires provinces. As Major explains:

> Each of the five riverine provinces has interests somewhat different from those of the others, and from those of the national government. Since some of the riverine provinces or some areas within them have few resources aside from the river; given the historic importance of irrigation to many areas in Argentina; and given the plans that the separate provinces have for development that would if all brought to fruition require water in excess of the capacity of the river, the decision problem is of great practical as well as theoretical interest.[4]

Needs of the sparsely populated provinces for water-control and irrigation projects conflicted with the preference of more populous ones for industrial electricity. Similarly, priority sites for certain irrigation installations implied depriving others of sufficient volumes of water for irrigation elsewhere in the river system. The MIT Argentine team sought to join multiple optimality (the combination of economic, jurisdictional, political, and social benefits) to hydrologic feasibility. All officials interviewed, as well as written documents bearing on the project, emphasized the role of Argentine officials in the project's design. The training component of the project was meant to give Argentina a team of six young professionals committed to working for Argentina's water resources agency for three years after returning home from MIT. This team would, ideally, not only utilize the new methodology to make practical decisions about the Rio Colorado but would also adapt it to water-basin development throughout Argentina. My interviews unearthed no fundamental or basic disagreements among interested parties.[5] All agreed that the three objectives of the project had been met. Criticism, freely expressed, focused on procedural difficulties encountered in carrying out joint actions. Nevertheless, clear divergences existed among the parties in terms of the relative priorities they assigned to the three common goals. Moreover, in discussions with MIT project officials, questions of value conflict were not answered directly or convincingly.

Tensions and Procedural Defects

Initial expectations diverged. Because the river is not navigable, Argentina's national government has no jurisdiction over the Rio Colorado (except in the case of navigable waterways, Argentine law assigns jurisdiction to individual provinces). One government official explained that investment decisions for the Rio Colorado had been pending for more than twenty years; no effective solution to conflicting claims on investment, placement of dams, and arbitration among parties desiring irreconcilable water uses could be found. Another official, himself the son of a former governor of Mendoza Province, was eager to remove any hint of political favoritism from his proposed solution to the impasse. Thus he decided to call in a prestigious US university to achieve his aims, while declaring that the "technical advisability" of MIT's final recommendation would reduce the danger of adopting a purely "political" solution. For public-relations reasons the project was "sold" to the Commission of the Provinces as the way to solve the Rio Colorado's practical difficulties, although within national government agencies it was asserted that the main benefit from the contemplated "technology transfer" would be the training of a sophisticated Argentine team. A loan of $380,000 from the Inter-American Development Bank to the Argentine National Fund for Pre-Investment Studies provided funds for the initial phase of the contract with MIT. An important personal element intervened: the cabinet-level officer entrusted with the decision was himself a water expert and had worked at the United Nations with one of the MIT engineers. The original contract stipulated that the sum of $380,000 was to be paid to MIT for the first two years' work.

The hierarchy of relative priorities among the main actors in the project varied. For the Sub-Secretariat of Water Resources, the first priority was the training of an Argentine multidisciplinary team able to handle overall water-resource-planning issues; its second priority was the improvement of a methodology for engaging in such activities; and its third priority, obtaining practical investment recommendations for the Rio Colorado. MIT, however, had a different ranking of priorities: first came improvement of methodology; next, training an Argentine team; and a distant third, providing practical investment recommendations. For the governments of the five interested provinces, the order was: practical investment recommendations, training, and methodology. Most conflicts arose when one party judged the other to be ignoring, or giving insufficient attention to, its own first priority.

The general lesson is that although identical priority rankings are not essential to success, the degree of procedural friction is closely correlated to the degree of consonance in goal-priority rankings. This theory finds concrete expression in tensions between MIT and the water agency over the training and methodology goals. MIT attached great importance to perfecting its methodological instrument, mainly because it was vigorously seeking contracts in other countries. This led one senior Argentine official to complain that "MIT did not transfer the technology: it formed it and perfected it in Argentina, thanks to our laboratory." Given MIT's priority scale, Argentines felt at times that insufficient attention was being given to their training needs at several levels. Although several Argentines suggested that it would have been better to bring MIT trainers to teach the team locally, trainees themselves disagreed with this opinion. At the same time, however, Argentine students at MIT complained of not being treated as regular master's degree candidates and of not receiving training that was specifically related to their future needs. Worse still, seminars staged by MIT at Neuquén and other Argentine sites produced disappointing results because MIT cast its teaching in purely hypothetical terms (around a fictitious Rio Tinto case) and refused to answer questions posed by provincial personnel about the real Rio Colorado. More than twenty-five MIT personnel were shuttled to Argentina, many of them professors or graduate students floating within what one Argentine called a "cultural vacuum; they knew nothing of local history, culture, psychology, institutions, or constraints." Perhaps because of this failing, MIT "experts," in their training efforts, repeatedly shied away from addressing the difficult *political* elements which, by definition, should have been included in multidimensional planning of river systems, because it was precisely such political elements which had proved so difficult for Argentines to handle and had moved them to summon MIT for help. Other failings are traceable to changes in top personnel, both at MIT and at Argentine host institutions.

On balance, the agreement reached by the five provinces (in December 1974) to a "certain configuration" (that is, location and nature of dam sites) of the Rio Colorado investment scheme stands as an undeniable step forward. And upon their return, the trainees were well equipped to handle planning for Argentina's overall water problems. The real long-term difficulty, according to one trainee, is how to raise the general level of expertise of the 6,000 engineers in Argentina. A major obstacle is the lack of solid information. Consequently, the Institute of Applied Science and Hydraulic Technology, whose research program he now directs, plans to create an information bank on natural resources. He explains that the country holds one hundred years' worth of nonprocessed information and that it will take at least five years to process relevant data. The most vital lesson he learned, he adds, is this: Argentina's ability to negotiate sound technology transfer contracts is tightly conditioned by its capacity to analyze relevant data.

MIT's Methodological Claims

Discrepancies arise between claims made by MIT experts and their actual performance in this first test of their methodology. One major problem is the way in which noneconomic factors are handled by MIT in its plural-objective planning model, an issue important to all planners who seek to quantify planning-input factors. MIT's treatment of nontechnical input factors reveals much concerning "trade-offs" among competing objectives of a project. One senior Argentine official declared that "MIT dealt with noneconomic inputs successfully in a qualitative way but did not succeed in treating them successfully quantitatively speaking." That is, although MIT paid great attention to these factors, it proved unable to express them quantitatively or to incorporate them organically into its simulations. When queried on this point, the MIT team leader replied that his experts made no attempt to treat social, political, or value problems (as distinct from technical and economic problems) as *inputs* into simulation or model runs. Instead, MIT tried to measure (quantitatively—but by what criteria?) what impact on the political, social, or value universe different hypothetical *outputs* would have. He confessed ignorance as to whether they had succeeded in doing this. In my judgment, MIT failed at this level, in great measure, because its experts suffered from "cultural vacuity," particularly regarding political culture. Notwithstanding the expressed disappointment of top-level Argentines over MIT's failure to quantify noneconomic variables, the university's scholars insist that the difference between quantitative and qualitative measures is meaningless. In the words of one MIT expert, "Everything can be measured in some way, and everything is quantifiable—some things with greater, some with lesser, precision."

His reply raises the question whether any foreign technical team can deal seriously with values as *inputs* and not merely as hypothetically projected impingement effects or imagined *outputs.* Perhaps value-input can be managed only by an indigenous team enjoying a solid mandate from the local populace which is the intended beneficiary of the technology transfer in question. MIT engineers are predictably skeptical on this point. Nevertheless, it is plausible to think that the ability of technical experts properly to assess value elements in plural-objective planning depends closely upon their degree of dialogue (in reciprocity—hence the need for legitimacy or mandate) with genuine representatives of the local populace. Nothing conclusive can be deduced from the Rio Colorado case, but value conflicts between promises and performance suggest that the hypothesis just outlined merits serious testing by those who profess interest in multi-objective planning.

This view is confirmed obliquely by the opinions of MIT professors who reported on their preferred criteria for site selection for new contracts using the methodology perfected at Rio Colorado. They prefer to work in a country where they are certain to find a high degree of discipline, professionalism, order, and willingness to work. Thus they were enthusiastic about Korea, pessimistic over the Philippines. And why? Because, notwithstanding their declared willingness to work in nonoptimum conditions (in such places, they indicated, as Sahelian Africa) for purely "humanitarian" considerations, they preferred to work most of the time where "results" had an "optimum" chance of occurring. This means places where the "objective conditions" for the applicability of their methods are in place: a unified command in water-agency decision-making within an agency that knows exactly what it wants and is willing to let the foreign consultant firm act according to its technological and professional exigencies. A rather strange requirement for a unit that insists on the ability of its model to incorporate social, political, and psychological factors in its *multi-objective* model. And all of this notwithstanding MIT's claim to have an instrument of transferable technology suited to less-developed countries.

On balance, then, it is clear that one must introduce some qualifications to Professor Major's conclusion that "while it is too early to say definitely, it appears that the MIT-Argentina project may well constitute a successful transfer."[6] One Argentine consultant thinks that one "must wait five years in order to gauge the success of the MIT effort at technology transfer." Perhaps so, but we need not wait that long to discover wherein lie recurring sources of value conflict between providers and users of technology. This case study identifies several such sources, even though the transfer on which it has focused is generally lauded, albeit tentatively, as a "success story."

Case 2: Precision Instruments in Latin America

This case illustrates how one reputable company dedicated exclusively to the manufacture of precision instruments transfers technology to its affiliates and clients. The Foxboro Company employs some 8,000 persons and manufactures approximately 1,000 products. Roughly half of its annual sales of $140 million come from overseas business, with 20% of total sales in the Third World. Foxboro, which specializes in systems and product technologies, makes precision instruments used to measure temperature, pressure, and flows of all types in operations ranging from copper mining to oil refining and food processing. Most of its "technology transfers" take place directly from the central manufacturing plant in Massachusetts to factory and processing sites around the world. For Foxboro, the key to satisfied customers is providing reliable technical services through the ongoing exchange of instructional documents, access to training facilities, and rapid repair and maintenance.

Foxboro is a well-established, traditional, and low-key company whose top managers are mainly engineers by training and managers by experience. The firm takes special pride in its ability to design, manufacture, and service the most complete line of instruments and systems available to the process industries. Products range from simple temperature gauges to sophisticated analog and digital computer-control systems. The approach to technology transfer adopted by the company seems quite congenial to the requirements of Latin American, and other, less-developed countries.

Facts and interpretations presented here are based on numerous visits to the main plant and R&D installations, coupled with frequent interviews with engineers and other officers at the main plant and at Foxboro facilities in Brazil, Argentina, Chile, and Peru.[7] This case study reveals the criteria used by one particularly responsible "seller" of technology.

Foxboro has been forced at times into measures it did not greatly desire, such as buying a manufacturing plant in Argentina instead of Brazil, its first choice in South America (it already has a plant in Mexico) and a more logical site. But, in the words of one senior company official, "One must sometimes do that sort of thing, especially when competition forces you into action." "Competition" is provided by Honeywell, Taylor, Kent, Fisher/Parker, Bristol, Hartmann-Brown, and Siemens. The wholly-owned subsidiary is Foxboro's preferred mode of association, although company policy dictates hiring as many local people as possible. A country's growth potential in large process industries is the key criterion governing entry by Foxboro into a national market. Because the firm sells instruments to producers, and not final goods to customers, it must be constantly alert to any source of demand: large industries with needs for many instruments, small industries requiring high degrees of precision, and state firms (particularly in mines, oil refineries, and steel mills) requiring specialized control systems. As with most firms with head offices located in the United States, Foxboro carries on the major part of its research and development activity at home, although laboratories are also located in England and Holland. Some pure research is carried out continuously on problems of fluid flows, but major effort centers on perfecting existing products and on anticipating the future needs of process industries. A particularly tight link exists between selling, R&D, and production engineering. Indeed my several visits to the main factory (and to one subsidiary) confirmed the image of the engineer as factory worker. Foxboro designs its own manufacturing equipment and builds most of it itself. It habitually has recourse to international bidding and often wins, even when it is not low bidder, because of its reputation for quality. It also advertises widely in professional journals and takes part in fairs and expositions. Most of its clients, however, are recruited as the result of direct visits by company officials. The firm spends relatively little for commercial advertising, preferring to let its "superior products and unmatchable servicing" do its advertising. In dealing with the Third World, the company declares itself interested above all in hardware.[8] To cite one spokesman, "We're not concerned with patents so much. We patent our instruments only so that no one else can reproduce them, not so that we can license them." Nevertheless, the firm does sell "application patents": these are *ad hoc* sales to customers who buy a patent for some particular application of a precision instrument. Unlike many other TNCs, the Foxboro Company displays no interest in diversification: "We are not interested in owning manufacturing plants of other things."

Through which mechanisms does the company transfer its technology? Except for one licensing contract in Japan, the usual way is the physical shipment to affiliates or clients of microfilm containing technical drawings. In turn, manufacturing subsidiaries in the Third World send reports and samples to the head office as part of an informal routine, not to meet the requirements of any written contract. One experienced engineer in the head office explained that there are two schools of thought within the company as to the merits of inspections for quality control. The first view holds that overseas manufacturers will obey precise quality specifications without any control from the head office; the second view contends that products must be constantly checked, sampled, and controlled. The same person adds that "performance history over the years shows that both systems have worked." Nevertheless, company policy insists on "the same standards of design and quality regardless of the manufacturing sources."[9] Notwithstanding concessions made to local requirements, "the function, performance, and appearance of the product is not to deviate from the corporate design." So as to ensure conformity to corporate standards, "all designs shall be under the control of the Corporate Development and Engineering Department. . . . [P]arts made by the various manufacturing facilities are to be interchangeable at a modular level to be determined by Corporate Development and Engineering, Corporate Marketing and the appropriate production plants."

Foxboro's Argentine subsidiary pays royalties to the head office on equipment designed by the latter. Yet the plant also uses equipment not designed by Foxboro; on such machinery, obviously, no royalty payments to the head office are made.

Many of the company's dealings in Latin America are not with subsidiaries but with sales and service representatives working on a commission basis. The political context of technology transfers carried out in this mode is illuminated by a brief look at decisions taken during the Allende years in Chile. Although Allende assumed presidential office late in 1970, the Foxboro Company had maintained an ongoing sales and service operation in Chile since 1968. By late 1972, however, the company became convinced that it would have a difficult time making profits in Chile. The office manager of the Santiago operation lamented: "All new projects were wiped out, we lost a big contract, and US banks withdrew credits for Chilean state-owned firms, which were some of our best customers."[10] Nevertheless, the company decided to keep the Santiago office open "in the hope of better days in the future." An indication of advantages accruing even to representatives paid on commission is gleaned from what then ensued. Foxboro offered this Chilean national the choice of a job with the company in Brazil, Argentina, Jamaica, Venezuela, or the United States. Largely for personal reasons, however, the person in question moved to Lima, Peru, where he reactivated a sales-and-service operation which had been defunct since 1967. In view of the Peruvian government's ambitious plans for nationalizing private enterprises and expanding further investments, prospects in Peru seemed encouraging. This spokesman preferred to deal with state-owned enterprises over private firms because the former have a clearer mandate to negotiate with outsiders and can pressure national banks and other government agencies to get the specifics of contracts "moving" (these "specifics" including import licenses, authorizations to transfer foreign currency, and registry of technological contracts).

He recalled a trip that he had once made to the state-owned copper mine in Chiquimata, Chile, for the purpose of convincing the nationalized mine that it should continue to purchase its control instruments from Foxboro. This engineer-manager employed interesting arguments. Under discussion was the cancellation of orders from Foxboro and a contemplated switch to Siemens, a German competitor. The Foxboro representative argued that if it were true that the United States could control Chile through its transnational corporations, what was to stop Germany from doing likewise through its own companies? Moreover, how could Chileans working at nationalized mines be completely sure that ITT did not own stock in Siemens and would not welcome gaining, through that company, a different foothold in Chile once its telephone operations were expropriated? The implicit value revelation here made explicit by my interlocutor is that a country cannot counter dependency just by looking at appearances. To him, it made no difference if a regime was communist, socialist, or capitalist as long as his own, and his company's, liberty to operate were respected. The second ingredient of "harmonious technology transfer," he added, is the existence of unambiguous rules for bargaining and doing business. The precise formulations articulated here by one person reflect the general attitude of TNC personnel working in the Third World. Such people resent insinuations that they are tied to "capitalist" regimes: all they ask is "the freedom to do business according to clearly defined, and observed, rules."

A glimpse into Foxboro's flexibility in technology transfers is gained from a visit to a wholly-owned service-and-sales subsidiary in São Paulo, Brazil. The transfer process (from Foxboro/USA to clients who purchase instruments via the intermediary of Foxboro/Brazil) rests on a constant flow of documented instructions for assembling, operating, maintaining, and repairing precision instruments. Top-level engineers in most Brazilian process firms read English and therefore enjoy direct access to all the technology. For the benefit of technicians and skilled workers at the next lower level, however, Foxboro/Brazil conducts training sessions around four volumes of master instructions, updated constantly with new technical information and supplemented by glossaries of technical terms sent to engineers in relevant industries. Many instruction manuals have been translated into Portuguese. Moreover, the enlightened director of the Brazilian operation sought government approval for his training program as a credit-granting technological unit. He has also urged SENAI (National Industrial Apprenticeship Service) to send its pupils to his own course free of charge. Another modality of "technology transfer" said to benefit not only clients but also "the larger cause of Brazilian development" is the sponsorship by Foxboro of mobile courses, running from a few days to six weeks, for such entities as Petrobrás, the government petroleum monopoly. According to this Brazilian director, a manufacturing plant in Brazil had become (by early 1970) a necessity for Foxboro. The major contribution of a plant is not in manufacturing itself, he explained, but in improving the training of one's own manpower. To him technology transfer is "simply a question of economics. But it takes time and money to train manpower, and it can be done best in your own plant."

Because Foxboro depends on large process-industry investments, the size of its potential markets is severely limited. Its area manager for Latin America estimated in late 1973 that the Latin American market for precision instruments was approximately $35 million annually, of which Brazil would account for $15 million. At one time the company had captured 60% of the Chilean market of some $3 million annually and more than half of the Argentine market, then estimated at approximately $5 million per year. Therefore, in periods of stress or transition, what "carries" the company is often a contract with a single large state-owned enterprise, as was the case with YPF (Yacimientos Petrolíferos Fiscales) in Argentina and CODELCO (Corporación del Cobre) in Chile. One of the company's main selling points is that it provides something more than quality equipment or even servicing of that equipment. Especially in power industries (the firm has "instrumented" more than 500 power installations in the United States, Canada, South America, Europe, Asia, Africa, and Antarctica), Foxboro often assumes contractual responsibility for overall system performance. The company is especially proud of its power-oriented computer system, PEIR (Performance Evaluation and Information Reduction).

Even a summary profile of Foxboro's approach to technology transfer would be incomplete without mentioning the impact of even the slightest research improvements in its instruments. One highlight of my several visits to the home factory came when an engineer dismantled, in my presence, a liquid-pressure gauge. His gesture came in reply to my question, "What makes a technology competitive?" The technological "forward edge" in this instance consists of a metal diaphragm in the center of which a small quantity of liquid silicone has been inserted. The diaphragm and the entire gauge roll even under slight pressure changes. But although this silicone-filled diaphragm is the key to Foxboro's competitive position in this instrument, the firm has no patent on the diaphragm, for Foxboro's real lead is in a highly refined welding process which no competitor could duplicate in less than six months. And by that time Foxboro would already have made further incremental but significant gains in refining its welding process.

This example illustrates the "fluidity" of incremental technological improvements obtained from research. The lesson for Third World negotiators is that what Andean Pact specialists call "modular" technology is something dynamic, not static. Ultimately, only the ongoing capacity to register parallel incremental improvements can enable a "receiver" of technology to implement a policy of disaggregating technology packages into their component elements. This is probably the most significant conclusion to be gained from the Foxboro example which, to all appearances, is a reasonably successful technology transfer.

Case 3: Frozen Foods in Brazil

The present example illustrates the criteria of a well-known US consultant firm in diagnosing one specific set of technological problems at the request of the government of Brazil.[11]

In the case under review a final "operations" contract was never signed. Nevertheless, the preliminary study conducted by Arthur D. Little, Inc. (ADL), under contract to the Ministry of Planning of the Government of Brazil, is instructive on three counts:

(a) It brings to the surface the values of a prestigious international consulting firm.

(b) It explicates several assumptions as to development priorities held by the client, the Brazilian federal government.

(c) It raises broad questions as to the "appropriateness" of decisional technologies habitually favored by international consultant firms.

One reason for the Brazilian government's interest in the project was the desire of the Medici regime to publicize a large and sensational achievement before handing the presidency over to General Ernesto Geisel in early 1974.[12] Contract feelers were first tendered to Brazilian authorities in 1972 by ADL's Rio de Janeiro office. Food experts in the company's Cambridge, Massachusetts, office subsequently refined terms of the project. After the probable impact of a cold-chain food system upon broader socioeconomic activities was explained to them, Brazilian officials began to show interest in the study. These officials stated as their goals for the project: to promote export earnings, to engage in greater regional food distribution, to control inflation by gaining mastery over fluctuations in demand and supply of food, and to achieve greater income equalization (although they never explained how equalization could be achieved). The federal government also expressed an interest in building central installations where refrigerated and frozen foods could be stored, thereby reducing waste and controlling peaks of supply and demand.

The preliminary assessment made by ADL and published in the two-volume report cited in these pages required one month's work by a five-man team in Rio de Janeiro. The follow-up study recommended by ADL would have cost more than $700,000 and required fifteen months' additional work; it was never contracted.

As discussions began, both partners agreed that Brazilian consultants lacked the time, the experience in general-systems approaches, and the objectivity required to plan a comprehensive cold-chain system for the country and to assess its regional impact.

Inasmuch as the larger, second stage of the project was never implemented, I shall confine myself to analyzing elements of the preliminary study germane to the three points mentioned above. Afterwards I shall briefly assess ADL's operational style (transcending the scope of this single example) in conducting diagnostic activities which bear on technology transfers to the Third World.

ADL's preliminary report assessing Brazil's needs in a cold-chain food system was presented to the Ministry of Planning in August 1972. To date (June 1977) no decision has been taken in proceeding to the next step, a detailed feasibility study prior to implementation.

A cold-chain system (CCFS) is defined as

> that portion of the food-distribution process and infrastructure which reduces and maintains perishable commodities at lower than ambient temperatures from production up to and including storage with the final consumer. A CCFS can theoretically exist for each commodity, and an overall CCFS can theoretically exist for all perishable commodities.[13]

According to the ADL report, the rationale for arguing Brazil's need for a CCFS centers around the following general objectives:

(a) to reduce food loss through spoilage

(b) to encourage food production in areas where facilities to conserve food are presently lacking

(c) to provide greater flexibility in the distribution of perishable foods thanks to refrigeration and frozen-food transport capacity

(d) to create sound storage capacity necessary for storing surpluses so as to control fluctuations in demand and/or prices

(e) to enlarge opportunities for farm people to sell their products in distant markets

(f) to endow the country with the ability to compete in world exports

(g) to reduce public health hazards posed by spoiled or infested foodstuffs

(h) to improve nutrition in the national diet

The Brazilian government concurred in the view that these goals would bring clear benefits. ADL consultants adduced still further advantages to installing a nationwide cold-chain food system, claiming that developing a CCFS would:

(i) increase productivity in agriculture by increasing the demand for goods and services required for building and operating a cold-chain food system

(j) demonstrate to producers the value of improved technology and efficient management of resources

(k) reduce domestic demand for imported food products

(l) lead to long-term price reduction in some foods through more efficient handling

(m) stimulate wide distribution of income by bringing regions of Brazil now virtually outside the market economy directly into that economy

(These objectives are listed in the report under the rubric: "Cold

Chain Food System Would Contribute to Brazil's Development Program.")

Which value assumptions pertinent to development emerge from the report? The arguments used to convince the Brazilian government that it "needs" a CCFS illustrate the "vital nexus" among basic value options, preferred development strategies, and concrete policy (in this case, a policy for food conservation). One way to clarify value assumptions is to pose critical questions about declared goals. Another is to compare expressed objectives (either explicitly declared or revealed in interviews by negotiating parties) with detailed targets presented elsewhere in the report and cognate documents. A third is to evaluate a concrete case in the light of broader criteria, such as those proposed by Ivan Illich in his works on education and health.[14] Illich considers it counterdevelopmental to attempt to satisfy real human needs (like the need for education, health, or food) solely through the provision of specific packages of goods or services which are then symbolically presented to people as "the only way" or "the best way" to meet those needs. His rationale is that these proposed "packages" usually entail high social costs or exclude large numbers of "needy" people from effective access to the very goods which allegedly justify providing the packages in the first place. It is instructive to review briefly some implications of the CCFS project in this light.

No one can quarrel with the objective of reducing waste through spoilage or of introducing rationality in the processing, storage, and transport of foods of animal origin (meats, fish, eggs, milk products) and of perishable fruits and vegetables. Nor can one dispute the assertion that cold-food handling should be initiated at the source of food production or that

> the system should be integrated, with links between ice makers, shippers, truckers, other transporters having equipment for conservation of cold foods, cold storage facilities, processors, distributors and marketers of perishable products requiring cold storage and/or handling.[15]

But the vital question is: Who will benefit from all this infrastructure? We glimpse the answer when we are told by the consultants that "if the system is to be fully successful, single-family units should be equipped with refrigerators and freezers as well."[16] Whatever may be the subjective intentions of the consultants on this point, the design of a system whose full success presupposes the existence of family refrigerators and freezers automatically excludes from the pool of potential beneficiaries the poorest masses who suffer most from food spoilage but are unable to purchase refrigerators or freezers. How, then, can it plausibly be argued that the creation of an adequate cold-food chain will lead to the evening out of income distribution?[17] ADL officials queried on this point replied that the "evening out" of income they had in mind is geographical: agricultural regions would gain a relatively higher share of national product than before. But they make no attempt to analyze income-distribution effects of the CCFS on segments of the population within agricultural areas. Moreover, it is not evident how the design system would allow Brazil's agricultural poor to improve their diet or gain access to better foods. On the contrary, one can reasonably fear that an increasing proportion of resources available for food-growing, processing, and distribution will be pre-empted by that "modern" sector of the economy—now expanded to include a CCFS—which already places many basic goods and services out of reach of all except the more privileged sectors of the population. A bias in favor of meeting the wants of those with present or future purchasing power is thus implicit in the very technological diagnosis made of the problem. Moreover, incentives to production are weighted in favor of "quality" producers, a euphemism for middle farmers and large agribusiness firms. Thus we read that deficient cold-storage capacity for meats causes farmers and ranchers to suffer, especially "ranchers who work to develop a high-quality hog"; they cannot sell their hogs for a premium "because the distribution system cannot carry the premium quality forward to the consumer with certainty, because of lack of an adequate cold-chain food system."[18]

The language employed in the ADL report illustrates a general principle discussed in a later chapter: namely, that modern technologies have an innate tendency to favor the rich to the detriment of those in greater need. The fault is not traceable to lack of vision or social responsibility in Arthur D. Little's professional staff; it is inherent in the very technologies consultant firms are best trained to manage and transfer. Only the recognition by "technology receivers" in developing countries of the existence of this systemic bias can even lead them to question the social impact of such proposals.

The CCFS under discussion also favors large-scale investment and leaves unexplored the issue of whether smaller, decentralized applications of capital might prove more congenial to the professional goals of the project. After surveying more than 1,000 beef-slaughtering houses, ADL consultants discovered that fewer than 10% of them possessed modern refrigeration facilities, a deficiency directly related to the scale of units. More than 56% of the units slaughtered less than ten head per day, and only 12% had the capacity to slaughter more than 100 animals per day. The food experts concluded:

> Such small businesses cannot readily afford the fixed investment necessary to provide adequate chilling or freezing facilities; in the absence of legal action by the governments, they would seldom consider such an investment.[19]

Once again the assumption is made that large-scale operations are to be preferred over smaller ones. If this is so, it then becomes plausible, perhaps even unavoidable, to channel infrastructure investments in ways which favor large agribusiness units at the expense of small producers. Inasmuch as the Brazilian government likewise endorses this outlook, ADL judges that, contractually speaking, it is meeting its client's needs.[20] The relevant point is that the choice of diagnostic technology often prejudices outcomes. Throughout its report ADL places exclusive emphasis on high technology, as when we are told that a CCFS "will provide a strong impetus for high technology cattle production in areas more removed from consuming centers."[21]

In their efforts to "sell" the complete cold-chain food system to their client, the ADL consultants paid scant heed to the needs of poor rural masses. They apparently gave no thought to the possibility that partial cold-chain systems adapted to local crops and purchasing power might prove more appropriate. Moreover, the report emphasizes production for world markets, arguing that more meat must be produced in order to meet export demand. Brazil's dearth of international-quality export facilities for frozen foods is cited as proof that the country "needs" a CCFS; nevertheless, elsewhere in the report it is acknowledged that equipment in cold-chain units is "difficult to maintain" when it is of foreign manufacture.[22] The consultants also flatly declare that more meat should be consumed by Brazilians,[23] offering no analysis of relative tradeoffs between acreage planted with grain to be used for animal feed and acreage devoted to crops allowing human consumption of protein lower on the food chain. Still another important value is implicitly endorsed in the statement that the frozen- and refrigerated-food infrastructure is a "subsystem of the larger agribusiness (or agri-industrial) system."[24] The appropriateness of a CCFS is thus justified by virtue of its compatibility as part and parcel of a larger system: it "interfaces with the international market, and with the durable and non-durable service sectors of the general economy."[25]

Notwithstanding the claim, noted earlier, that a CCFS would reduce Brazil's need to import food, the report takes it for granted that "imported refrigerated and frozen foods leave the CCFS from many points in the system."[26] Nowhere is the report more questionable, however, than in its claim that the CCFS will contribute to income equalization, judged desirable because "inflation has a more severe effect on lower income groups."[27] One cannot but be skeptical of this assertion in a document totally oriented toward high purchasing power—as when the client is told that it must prepare for expected demand for "TV dinners or other *important* frozen food items."[28]

What emerges clearly is the conclusion that even responsible consultant firms such as Arthur D. Little—whose top leadership has a genuine social conscience at the international level and whose self-image is that of an enlightened, tolerant company where bright people have great freedom to be creative[29]—do not carefully scrutinize the larger value implications of international consulting. Although they locate consulting at the "cutting edge" of developmental activities, in practice, according to one ADL official, their predisposition is simply "to see if we can do a job for clients who have money to pay."

Many sensitive consultants are aware of discrepancies between the moralistic rhetoric of "helping" underdeveloped countries achieve their genuine goals and the commercial reductionism of their dealings with government agencies or private business in these countries. But such value tensions as those brought to light in this cold-chain case seem to be a natural outgrowth of the manner in which consultants compete to transfer their diagnostic and prescriptive technologies to the Third World.

Notwithstanding these discrepancies, which ADL openly acknowledges, the firm remains optimistic about the future evolution of relations between consultant firms and less-developed countries. Company leaders favor regulation—largely self-imposed—of transnational corporations to make them more responsive to legitimate social pressures. And ADL is confident in its ability to stay in the forefront and avoid what it calls "pedestrian" technology contracts. One basis for its optimism is the firm's strength in "management technology," the application of which opens "tremendous opportunities in many countries." The real problem here, the company explains, is to shorten the time gap between the discovery of a new technology and its application. So as to reduce this gap, ADL devotes much energy to the marketing of technology. In the race to market, however, consultants testify that they cannot indulge in the luxury of questioning the values of their clients beyond the point of assuring themselves of two conditions: that the work requested serves honest ends and that professionals can engage in it without betraying their code of professional integrity.

This case study of the cold-chain food system suggests, however, that vital systemic value conflicts can easily be overlooked if these two principles are applied in isolation from wider norms of social responsibility. (To restate an earlier point, there are many important social "externalities" that are never "internalized" in the process of transferring technology.) ADL is keenly aware of this danger when it evaluates the behavior of individual enterpreneurs in a client country. While reviewing trends in private enterprise within Brazil, for example, ADL experts detected much dynamism, as many firms were building new cold-chain food units. But although these innovators are to be commended, the ADL report adds, "their prime interest is the financial future of their enterprises; they have limited reason for concern about the technological coherence of the system as a whole."[30] One must turn ADL's evaluation back on ADL itself and ask: Why are you unconcerned with the coherence of Brazil's developmental system as a whole?

My argument, in short, is that even such a laudable goal as "technological coherence'' of the system is too narrow a framework within which to transfer decisional technology. The vital nexus requires that technological coherence be linked to development strategy and the basic value options of the society in question. The cold-chain study suggests how difficult is this task.

Case 4: Tourism, Technology, and Values

Unlike those preceding it, the present case study bears on the impact of technology not in a specific project but in one *sector* of activity. The following pages highlight value dilemmas posed by technology transfers in the tourist industry.

A wide array of technologies is used by promoters of international tourist activities.[31] These include transport technologies, public-relations techniques, image technologies (films on tourist sites; special cable, radio, and mail installations; etcetera), construction technologies (for hotels, restaurants, amusement centers, holiday villages, resort installations of all sorts, recreational infrastructure), management technologies, financing technologies, and recreational technologies (for special facilities like marinas, golf courses, swimming pools and for special functions such as organized visits to archeological sites). Food and cold-chain technologies also figure prominently as adapted to supplying tourists with ''international quality'' food and refrigeration.

No single technology, however, is so important to tourism as the intangible skills of fantasy creation, a specialization which the French cultural historian André Malraux claims characterizes Western modern civilizations.[32] The public in rich countries is massaged, with the help of multiple technologies, with images designed to induce it to spend money on tourism, preferably in poorer countries. Happiness is surf, sex, and sand. Alternative fantasy-creation takes the form of reducing culture, history, religion, and archeology to bring consumer *objects* rather than internalized *subjective* enrichments. Through the bias of image manipulation, promoters of tourism give a content to the ''notion of desirable development'' for the populace in host countries. Tourism, more than others, is one investment sector wherein value considerations cannot remain externalized with impunity; they must be internalized. The problem has often been ignored, even by ''experts.'' To illustrate, World Bank specialists, in a 1972 document, defend their policy of employing

> the same criteria in evaluating a tourism project as in evaluating a project in, for example, agriculture, mining or manufacturing. A tourism project is considered appropriate for Bank financing when the economic rate of return is at least equal to the opportunity cost of capital in the country in which the project is located.[33]

This purely economic approach does not lead to the choice of a tourism policy supporting sound development, a fact acknowledged by recent World Bank documents. The report of the Inter-American Development Bank, on the other hand, is sensitive to these problems. We are told therein that tourism brings its own evils and that three special problems concern tourism in South America:

(1) In small island economies in the Caribbean, the net social benefits of present patterns of tourist developments are exceedingly small.

(2) Disruption by large-scale tourism of the economic functions and structures of smaller-scale economies is substantial.

(3) Generally tourism is more capital-intensive and more generative of import demand than has been thought the case in the past.[34]

That all is not well even when tourism is "successful" is also suggested by a study, published by the Organisation for Economic Co-operation and Development, in which governments are urged to diversify the economy of rural areas by promoting "rural tourism"; to imbue tourism policies with a "social content" (protecting consumers, increasing the accessibility of wider sectors of a population to recreational facilities, and conserving natural and cultural beauty); and to grant the public a role in planning tourism so as to protect its interests.[35] No industry caters so blatantly to the wealthy and middle classes as does tourism. Worse still, it strives mightily to induce more modest spenders to convince themselves that they too can afford "luxury" vacations. Most promotional and analytical literature stresses large-scale, mass tourism with little regard for equitable access or larger issues of social justice.[36]

What, then, are the arguments for a country's investment in tourism? First and foremost is the proposition, expounded by lending agencies and consultant firms even in poor countries, that tourism is a beneficial source of foreign currency. Superficially, this may be true, but such income is subject to immediate drainoff through numerous leakages. Among leakages identified in the Inter-American Development Bank study are expenditures for imported goods and services consumed by foreign tourists (most tourist promotion creates or reinforces the "needs" of foreign tourists for imported goods), payments of interest and amortization of foreign capital, payments to expatriate workers, costs for training abroad, and imports of capital goods for the tourism sector.[37] A more intangible cost is the pressure placed on poor local populations to imitate the consumer behavior of tourists, thereby generating new levels of local demand for imported goods. For these reasons, the *net* foreign earnings from international tourism are sometimes less than 45% of *gross* foreign-exchange earnings.[38]

The second argument invoked to justify tourist investment in poor countries is that it creates jobs. But, if we are to believe the World Bank report,

> even for many developing countries where tourism has become a leading foreign exchange earner, the sector's output constitutes a

relatively small portion of the GNP and employs directly only a small part of the labor force. It is often claimed that tourism is relatively labor-intensive but the available evidence is not conclusive on this point.[39]

Moreover, there is something particularly shocking about luxurious installations in locales of mass misery. Recognition of this scandal has led many governments to seek ways of "integrating" social remedial investments with their "development" of tourist resorts. One proposal describes the imbalance between luxury tourism and generalized squalor in these terms:

> The development of Acapulco as a tourist center and as an urban and regional community has not been balanced. It is estimated that of 175,000 inhabitants of the port, 105,000 live in low-income neighborhoods which are largely without adequate public and municipal services. The contrast between the low-income sections and the milieu in which tourist activities take place has become more striking in recent years, primarily as a result of the rise in the economic status of tourists and of migration to the city from surrounding rural areas. The rapid expansion of the tourism sector and the growth of the low-income population threaten to create a situation of conflict.
>
> The coexistence of tourist zones with depressed areas of the city and the region could give rise to social frictions and even to curtailment of the inflow of tourists, with effects on the regional and national economy.[40]

There is no need here to detail the complex maneuvers which ensued; briefly, the Mexican government agency in question negotiated several alternative contract modalities with US consultants, at first with proposed World Bank financing, later without it. The point is that Mexico's government chose to ignore structural imbalances resulting from a defective tourist policy and to deal merely with symptoms. Tourism revenue in Acapulco had dropped rapidly because the bay was being polluted by open sewage systems. But for political reasons this was not acknowledged publicly because Miguel Alemán, a former president of Mexico and now "tsar" of tourism in his country, owned extensive tourist properties in Acapulco.

This type of conflict between developmental values and tourist technology—at planning and managerial levels—has led some tourist professionals to plead for a "new tourism" designed to promote the development of the populace at tourist sites. This interesting movement has made some inroads in the Caribbean area. Its principal theorist is Herbert Hiller, whose objective is "to resolve the contradictions between tourism and development,...to ask in what way tourism can be supportive of development."[41] Although tourism investment in poor countries is presented as an aid to development, an initial contradiction is apparent in the fact that tourism promotes the values and technologies of only the industrial rich world and its leisure classes. For Hiller a second value conflict lies in the inability of the general populace at tourist sites to control the tourism flow; for, after all, "progress" now depends on the affluence and leisure of tourists from other lands. A third difficulty arises from the apparatus mounted by the tourism industry in order to "industrialize" the leisure of tourists in marketable ways.

In positive terms, Hiller urges placing the development of the people of host tourist sites at the heart of the tourist equation. How can the people's objectives be met? he asks. By what kind of tourism on what scale, in what patterns? Priority must go to these objectives: optimizing local self-sufficiency, utilizing trade (including tourism) to increase domestic benefits from local resources, and defending local culture as a valid expression of adaptation to natural resources and constraints. In his words,

> The objectives of development will include establishment of institutions and symbols of cultural adaptation to the resource environment, the integrity of local communities, the investment of our lives in purposes locally sanctioned.

Ultimately,

> The success of tourism will be measured by how well these and related objectives are supported through the energies of the local community in organizing for the presence of visitors.

Hiller's specific proposals include: people-to-people programs; the creation of local and national tourism cooperatives; the maximum use of local products in accord with local tastes; the encouragement of locally scaled businesses through direct contact between craftsmen and visitors; the fostering of tourism in rural areas; the provision of tourism-related training programs at community and national levels; measures to exclude tourism from communities not wanting it; the preservation and improvement of historical sites; and the dispersal of visitor activities throughout broad reaches of the community.

"New tourism" calls for marketing strategies which focus on the quality, not merely the quantity, of visitors. To increase the real income of host *populations* (not simply to fill the coffers of host *governments*) becomes a major objective. Hiller encourages hospitality toward certain categories of visitors who would contribute to understanding between their cultures and that of host countries: students, minority groups, emigrants from the host country, persons with occupational or hobby linkages to the receptor countries, and educators. Much of Hiller's work aims at changing images among travel-marketing professionals of "what tourists want" and at supporting efforts by tourist-dependent societies, particularly in the Caribbean, to institute new tourism policies which serve local interests.[42] An eloquent statement of these aspirations comes from the former premier of the unspoiled Caribbean island of St. Vincent, James F. Mitchell, who wrote in 1973:

> As Premier of my state, you will pardon me, I hope, if I appear not too anxious to grab the easiest dollar. The tourist dollar alone, unrestricted, is not worth the devastation of my people. A country where the people have lost their soul is no longer a country—and not worth visiting.[43]

Nowhere is the "inappropriateness" of mass-scale market technologies more apparent than in tourism. This is why Hiller wants to replace the "high-technology hotel" with other forms of construction and services which support the development of poor lands heavily dependent upon tourism. Little evidence exists, however, either in official publications or in the reports of private consultant firms, that tourist technologies and marketing procedures are being subordinated to the properly *developmental* needs of host countries or even of industrialized nations with "export" tourists. In the hope of introducing correctives to bankrupt philosophies of tourism, the "new tourism" school analyzes the benefits accruing to tourists themselves when they have a more genuine, development-fostering experience with the people whose lands they visit. "New tourism" obviously emphasizes the values of local cultures—viewed not statically but in a self-defined developmental dynamism. Yet the true leisure needs of tourists themselves are seen to depend on respect for the hosts. This emphasis stands in marked contrast to the position of "leisure scientists" like Max Kaplan and their patrons, who concentrate on experimenting with "leisure communities" for the rich in the hope of finding new paradigms of a "humanizing utilization" of leisure time.[44] As Veblen, Pieper, de Grazia and Huizinga long ago pointed out, leisure has been the privilege of the rich.[45] Nevertheless, their consumption and symbolic patterns largely set the style for less opulent classes. Mass tourism, thanks to the technology it employs and the values it channels, is rapidly making all forms other than mass-consumer models of development nonviable in countless small and vulnerable societies. In fact, as presently conducted and financed by most international development agencies, tourism actually institutionalizes several *counter*developmental trends, among them:

- excessive dependence on outside capital
- a division of labor which casts nationals in menial jobs and foreigners in loftier management positions
- an excessive reliance on imported "international quality" goods and services
- the pre-empting of attractive natural resources for aliens, to the frequent exclusion of nationals
- the over-commitment of limited host government funds to providing tourism infrastructure, at the expense of vital services to the needy local population

- legislation favoring foreign ownership of tourist facilities
- the trivialization of cultures and peoples by tourist "images" which, as manipulated by promotional technologies, emphasize superficial delights in ways damaging to local identity and dignity[46]

Increasingly, however, host governments are beginning to alert themselves to the excessive value sacrifices they are making when they accept technology transfers on the terms of the international tourist merchandisers. And some of them are taking steps to devise alternatives. More and more people in the Third World are coming to recognize tourism as "poison in a luxury package."[47] The chief merit of the "new tourism" briefly profiled in these pages resides in its practical efforts to show that tourism need not be thus. The choice for poor countries endowed with tourist attractions is not: Either repudiate tourism or sell out your culture. Instead the lesson is: Promote a new form of tourism which is both locally developmental and humanly enriching for outside tourists.

Miscellaneous Short Cases

Widely differing circumstances, preferred operating styles of individual companies, and technical constraints within each branch or sector of industry all condition modes of technology transfer. In addition, varying degrees of stability in technologies themselves also constitute a major variable in transfers. Although exact coefficients of stability cannot be assigned to specific technologies, practitioners agree that some technologies are relatively stable, others highly volatile. The importance of varying stability in technologies is illustrated in the next two case studies.

Among firms visited by the author, ASTARSA (Astilleros Argentinos Rio de la Plata, S.A.), an Argentine shipbuilder, stands at one end of the scale—that of stable technology—whereas the Cabot Corporation, a US manufacturer of carbon black, deals in unstable technology.

A. Stable Technology: Dredges

ASTARSA, the largest private shipbuilder in Argentina, has, since its inception in 1927, built more than 130 ships, ranging from tankers to auto/passenger ferries and specialized cattle-carriers.[48] Other fabrication lines include pressure vessels for metallurgical industries, heavy machinery of all types, locomotives, earth-moving equipment, and army tanks. The company designs most of its own tooling machinery and remains technologically competitive thanks to a policy of diversified licensing with foreign firms.[49] Most ASTARSA licensing agreements cover just a few years, because the firm's own engineers, technicians, and skilled workers are not experienced enough to benefit fully from their training visits to the plants of their licensing partners.

Consequently, ASTARSA rarely needs to renew licenses once these expire. At present this firm, which employs some 1,500 people, holds licenses with General Motors, Caterpillar, Ellicott Machine Corporation, and M.W. Kellogg in the United States; Usines Schneider, Alsthom, Matériel de Traction Electrique, and Société Alsacienne de Constructions Mécaniques in France; Vickers and John Thompson in England; and Werkspoor in Holland.

Though it is primarily a shipbuilder, ASTARSA has diversified into earth-moving equipment, railroads, petrochemicals, military equipment, and metallurgy in order to offset oscillations in demand for naval construction which could lead to seasonal unemployment. The firm has a well-trained corps of workers and does not wish to see them unemployed during portions of the year. Its technicians have already assimilated most imported technology and are now able to comply with fabrication standards set in codes of the American Society for Mechanical Engineers, British Steel Standard, American Petroleum Institute, Interstate Commerce Commission, and Tubular Exchangers Manufacturers Association.

One of ASTARSA's licensors, the Ellicott Machine Corporation located in Baltimore, specializes in a form of technology which is highly stable, namely, the manufacture of dredges and dredging materials.[50] The selection, design, building, and maintenance of dredges is a highly specialized business requiring wide engineering experience and constantly varying applications in field work, design, production, and servicing. Each dredge must, in a sense, be "tailor-made." Ellicott, a traditional firm created in 1885, has representatives and licensees in seventeen Latin American countries. Its arrangements with ASTARSA incorporate several interesting features.

As background, it should be noted that although Ellicott favors licensing in general, it faces restrictive legislation in Argentina requiring that national products be used when available. Therefore, the company cannot sell its dredges ready-made. Even licensing poses problems because of high duty (100%) and the legal prohibition to import certain dredge parts (e.g., complete engines) normally purchased by Ellicott from General Motors and Caterpillar. Thus constrained, Ellicott in 1964 signed a licensing contract with ASTARSA (for five years and extendible thereafter) to build dredges. ASTARSA needed a license because, notwithstanding its capacity to build hulls and power systems, the company lacks the technology to build satisfactory winches, pumps, cutter assemblies, dustpan heads, and engines. The government prohibition on importing engines fabricated by General Motors and Caterpillar is neutralized by ASTARSA's commitment to the Argentine government that the relevant equipment will be taken out of the country once the dredging job is finished. (It is current practice in large jobs to shift dredges to other sites.) Interesting procedures are observed in bidding for jobs in Argentina: Local licensees are the prime bidders on government jobs, while outside licensors may contract with licensees to supply specifications and know-how, as well as a set of modules and winches.

Of interest to the present study is the relative ease and speed with which ASTARSA acquired a high degree of technological autonomy. The chief reason is that practically all the technology used is stable, that is, it changes slowly. Shipbuilding employs mainly product technologies embodied in tools and machinery, not in fluid processes. Safety and precision are the key variables, not packaging, consumer attractiveness, or ease of transportation. All these factors make for relative stability. And because all ASTARSA licenses include full visitation privileges to host plants, local capacity to improve upon licensed machinery and finished parts has developed rapidly. ASTARSA now builds all its ships with its own technology, with the sole exception of the know-how, covered by the Ellicott license, for the construction of special dredges. The Argentine shipbuilder's reasons for importing technology are reducible to two: (1) ASTARSA lacks the market volume to warrant developing its own technology (sales volume is especially vital in the production of capital goods), and (2) each of its ships must be especially designed and custom-made. Because specialized dredge technology had to be of the highest quality, recourse was had to Ellicott.

As a matter of general policy, ASTARSA's managers believe that, in cases of joint-equity participation, initial technology provided by foreign partners should be viewed as part of the investment. Consequently, payments should be made only for subsequent improvements. In the case of improvements made by local licensees, compensation should be made to them in the form of royalty payments by the original supplier of the technology. They also judge royalty payments, in general, to serve as counterincentives to inventive adaptations. This conviction explains why, in certain cases, ASTARSA has declined to renew a license; the company would rather stimulate its own personnel to find equivalent technological solutions. Overall, both ASTARSA and Ellicott expressed their satisfaction with the technology-transfer contract just outlined. The general lesson to be drawn is that such compatibility is quite easy to assure when the technologies concerned are relatively stable.

To round out the picture, it should be added that Ellicott conducts about 50% of its total business in underdeveloped countries. The company sells freely in Brazil, where no restrictive duty is in force and where import licenses are easily obtained. Although Brazilian legislation is similar to that in force in Argentina, the interpretation given by officials in Brazil is much looser. Ellicott also does a considerable business in Venezuela and Colombia but very little in other Andean Pact countries—Chile, Peru, Ecuador, and Bolivia. (One company official, while discussing the criteria adopted by the firm in its technology-transfer policy, explained that Ellicott has built nothing under license in Colombia because the skill level in that country is not yet sufficient for building dredges.) Generally speaking, however, dredging markets are growing rapidly in many parts of the Third World, especially in Latin America, where large projects are in progress in mining, dam construction, port modernization, river-navigation development, beach-resort improvements, and construction of new airports. The technological edge enjoyed by Ellicott resides largely in the quality of its dredge modules and the supporting electronic equipment used to control operations at each step of dredging. The company has pioneered a production meter which offers many benefits not previously available to the industry, such as direct readings of velocity and specific gravity of materials being pumped, instantaneous production in tons per hour, and total tonnage of material pumped. Ellicott has also introduced a new containerized portable dredge which greatly reduces transportation costs and mobilization-demobilization time. At its R&D site in Baltimore, Ellicott has facilities for simulating almost any conceivable problem environment. Notwithstanding the basic strength of its dredging operations, however, the company, like many others operating in stable technological sectors, has diversified. It presently has holdings in couplers for railroad cars and wheels for trucks and trailers, power-control equipment for nuclear generating plants and other facilities, and equipment for tension-stringing and construction.

The ASTARSA/Ellicott licensing agreement illustrates conditions under which successful technology transfers may take place.

B. Unstable Technology: Carbon Black

Founded in 1882, the Cabot Corporation had become by 1947 the largest producer of carbon black in the United States and by 1950 the largest in the world.[51] Like most large transnational firms, it has diversified and now derives its income from three main sources: performance chemicals (including carbon black), energy, and engineered products. These pages concentrate exclusively on carbon-black operations, wherein technology is subject to frequent and rapid changes. Of particular interest is the insistence of company officials on the dominant role played by technological leadership in maintaining a competitive edge.

Carbon black is obtained from a heavy, aromatic, residual fuel oil, with natural gas serving as a secondary source (or, as it is termed in the industry, "feedstock"). More than 90% of carbon black used goes to rubber applications. A tire for a passenger car contains six to seven pounds of black; an average truck tire, twenty pounds. Other uses include pigment in inks, paints, plastics, and paper. In addition to six manufacturing plants in the United States, Cabot has production units in Argentina, Colombia, England, Canada, France, Germany, Italy, and Spain. Company output of carbon black accounts for almost 25% of total world production, excluding socialist countries.

Technology faces several challenges in the carbon-black industry; one is to produce new, quality carbon black from what is called the "furnace" process, which allows manufacturers to phase out the "channel" process that is now becoming obsolete thanks to rising prices of natural gas. What appear to be minor technological improvements often lead to new products, specifically, varieties of carbon black with novel or improved applications. Cost competition is a third domain in which technological breakthroughs produce tangible competitive gains. Process technology (used to prepare the black) merges with product technology (the resultant black has different properties for reinforcing rubber or serving adhesive functions in nonrubber mixes). Cabot's research concentrates on extracting larger quantities of black per ton of feedstock, on finding additional uses for nonconventional feedstocks, and on synthesizing black from nontraditional processes. Pertinent to this study is the effect such volatile technology has on the mode of transfer operations to less-developed countries.

Company officials interviewed endorsed, unanimously, the view that wholly-owned subsidiaries are the preferred channel of technology transfer. Under this arrangement, "technology transfer becomes almost automatic, and questions of licenses and royalties become purely academic." Government pressure in several countries, however, has led to accommodations. Cabot, accordingly, now accepts joint ventures, holding 50% equity in Malaysia and Iran, 49% in Australia, 40% in The Netherlands, and 10% in Japan. Its Argentine and Colombian plants are wholly-owned, and the company contemplates building a new facility in São Paulo, Brazil. Cabot fears, however, that legislation imposing remittance ceilings in these countries will "eventually cramp the company's style." The new Brazilian venture will include less than a 50% equity for Cabot; the company insisted on this clause "in order to be able to charge a technical service fee to the Brazilian affiliate."

When asked their opinion regarding the aspiration of many Latin American countries to acquire their own research and development capacity, officials replied that it does not make economic sense for subsidiaries, or for poor countries, to build their own R&D installations; these are too expensive and scale does not justify investment. More importantly, Cabot wishes to maintain control over its own R&D. Having one's own laboratories allows one to plan ahead, to be the first to reap the benefits of technological breakthroughs (crucial in a "volatile technology" industry such as carbon black), and to assure access to technological innovation. In the absence of one's own R&D, competitors might choose not to sell the company and the new technology. Concessions had to be made in negotiations with Japan because of that country's huge market. Cabot licenses its technology there in two separate contracts: one for existing know-how, another for future know-how. Because carbon black is a "high specialty" product, it is subject to constant shifts in product quality. *But the key to quality is technology,* and therefore, control over technological change is the key to market advantage.

Cabot officials declared that there are three channels whereby less-developed countries may improve their basic bargaining position in technology:

- more demanding negotiation (as in Japan's case)
- tougher commercial terms for raw materials (in imitation of the OPEC countries)
- probes into new areas of technology development (for example, solar energy)

They claim that many undiscovered technological "points of leverage" exist which poorer countries could readily exploit. Although the company refuses to grant licenses to Eastern European countries because they insist on the right to sell in Western European markets, Cabot remains confident in its ability to adjust to changing demands from all types of governments. And notwithstanding its desire to retain technological control, it praises efforts by Brazil's National Institute for Industrial Property (INPI) to set up a computer data bank on technology.[52]

The Cabot Corporation exemplifies the competitive, albeit urbane, sophisticated, and "socially responsible," international company. Opinions of its officers here recorded, although personal and not necessarily reflective of company policy, are nonetheless confirmed by my observation of company practice. They suggest some correlation between the degree of stability in a technology and the ease with which licensing arrangements can be reached with host countries. They also imply that new ground rules for negotiation are possible whenever weaker partners utilize cost gains realized by scale production to invest in new technologies. Although volatile or unstable technologies may be more highly competitive than stable ones, minor gains realized therein can be more quickly capitalized in a broader market. This explains why the company searches for greater flexibility in exploiting such gains. To facilitate the task, the company grants its two R&D laboratories, located in the United States and Great Britain, relative freedom to concentrate on problem-solving of their own choosing.

We are left with no doubt as to the intimate link between R&D and marketing strategy. And control over technological change is more vital, in the long run, than short-term profits generated by diffused technological licensing.

C. Building Up R&D Capacity: The Case of USM

Third World governments seek not only to control technology transfers from the rich world but also to identify how competitive research facilities are set up. One US corporation, USM,[53] illustrates how a large R&D installation can be created thanks to the convergence of several factors: the vision and perseverance of company officials, unusual circumstances (in this case created by World War II), and a period of "learning by doing," which holds interesting lessons as to the alleged difficulty of new technology.

Long before R&D became a corporate byword, USM had achieved leadership in private industrial research.[54] One farsighted official in the company had built up, by the late 1930s, a team of 400 people engaged in research related to the company's sole product line, shoe machinery. This official, nevertheless, was convinced that a one-product company could not long survive, and he began preparing for future diversification.

When World War II erupted, United Shoe Machinery's research director, so as to avoid losing those he called "his bright young men" to the military draft, turned over to the US government his entire research installations and team. The armed services, along with other government agencies, accepted the offer. The research team, then numbering 500 people, later peaked at 720. Working under contract, the team studied everything from gun mounts for B-29s to anti-aircraft computers, solid-fuel rockets, control systems for torpedoes, gyroscopes, and wind tunnels. In the words of one engineer: "Our ignorance proved to be a great asset. We were forced to take apart computers and other pieces of equipment which we knew nothing about; to learn what made them tick; reconstruct them; and design improvements to solve the problems laid at our doorstep. Our team of eager-beaver kids started from scratch, played around with complex problems like light spectrums and radiation. Although this kind of research was over their heads, they quickly learned that solid basic research conducted by the Massachusetts Institute of Technology would help them. They learned when they had to."

The speed with which this team mastered intricate technologies outside its specialized fields under the pressure of direct problem-solving in a climate of incentives based on "helping the nation" is noteworthy. No less instructive is the decision taken by the R&D unit, after the war, to refuse further government contracts and concentrate on special problems faced by the parent corporation. USM researchers noted that they had not done any work for several years on their own industry, shoe machinery. On the other hand—and despite this lapse in development—the company's retention of a virtual monopoly in leasing shoe-manufacturing equipment made it increasingly vulnerable to a protraction of its long history of being "taken to court" on antitrust suits. The priority task was obvious: to diversify the company. By 1955 top managers had become sympathetic to this idea because they had had to sell off much of their centralized operation. And so, as part of its diversification strategy, necessary in order to survive and remain profitable, management decentralized control and acquisition in accord with technological R&D breakthrough capacity.

The process, although finally successful, proved difficult: even after research operations were organizationally separated from development, it took years to move away from prototype development to general-market production. The "long and difficult road" to viability included a decision, reached after much debate, to decentralize corporate research itself and to create separate laboratories for each of the company's major product divisions: machinery, adhesives, and fasteners.[55] Yet, today, a single senior research officer coordinates all efforts, "cross-fertilizes" the laboratories, and links separate group priorities to overall corporate decision-makers. The firm's 1973 annual report speaks of

> a degree of synergy in the group's operations wherein a machine may be developed in one location, the technology shared with the rest of the organization, manufacturing takes place wherever optimum quantities can be produced most efficiently, and the end product marketed wherever in the world the demand and the opportunity exist.[56]

The company sees the "emergence of Latin America as an economic entity" and the "stirring of China and the opening of its economic borders" as promising signs that its decentralized R&D policy, allied to a "global approach" of coordinated marketing, will be amply vindicated.

Company officers leave no doubt that technology is the source of their competitive edge. The greatest edge belongs to multitechnology companies able to eliminate obsolete technology lines and create new ones quickly. In their view, used technologies are highly appropriate in many less-developed countries, but their introduction is resisted by politicians for extrinsic reasons. Technology exchange with competitors and clients is like a chess game: "One must be in touch with opponents, but not too closely. 'Keep them guessing' is the watchword." Their advice to policy-makers in less-developed countries reads: "There is no way of stopping technology transfer. Perhaps you can control these transfers. But if you cannot, don't try to stop them. Instead, concentrate your efforts on finding ways of benefiting from them."

USM experience is interesting on three counts:

- It illustrates the multidimensional potentialities of having a basic research infrastructure, particularly its capacity to acquire mastery of unknown problem areas by trial and error.

- It points to the value, within the firm, of the "Sabato triangle" strategy—linking policy-makers dynamically with producers and researchers.
- It confirms the dominant role played by shifting technologies in the marketing strategies of a large, transnational corporation.

The company is more articulate than most as to its own role in a transnational world economy. We are told that productivity improvement is the major instrument for achieving economic growth, in these words:

> USM is convinced that productivity is the path by which the U.S. can best make itself competitive with low-labor-cost countries. ... Better productivity creates more jobs through real economic expansion, holds down inflation and enables high-labor-cost countries to compete with low-labor-cost nations.[57]

D. "Appropriate" Technology for Poor Peasants

In the Alto Valle (Upper Valley) region of central Bolivia, several Quechua peasant communities are experimenting with new modes of economic activity. Small villages clustered around Tiataco and Huayculi have adopted forms of producer cooperatives which depart in several important respects from conventional models.[58] Their approach to technology illustrates several important values germane to this study.

The economy of this dry plateau, located in the province of Cochabamba and the site of much armed violence in the Bolivian land reform of 1952, is based largely on subsistence agriculture around a protein-rich native crop known as *quinoa*. A few years ago, an indigenous movement, still of modest proportions, arose with the goal of diversifying sources of economic income in a manner which would help revitalize Quechua culture and self-identity. In the words of one of the movement's leaders:

> Cultural development of the people has two elements: the dynamization of the human potentialities and the cultural values of the community, and the assimilation of technology and science at the service of the cultural development of the people.[59]

The two villages just mentioned have launched two cooperatives: one to produce ceramics for sale, the other to make rugs, ponchos, and other marketable woolen artifacts. One broad objective is to improve the economic condition of the entire community, not merely that of members of the cooperative. This commitment to communal improvement helps explain certain decisions reached after arduous debate.

The first decision is that new technology will be judged "appropriate" only to the degree that the community at large is able to understand and control it. Specifically, the ceramics cooperative decided in December 1974 *not* to introduce small electrically powered kilns into the village. The background against which this decision was made is this: Traditional ovens use twigs and wood gathered locally for fuel, but such sources are now becoming scarce.[60] Moreover, this fuel produced uneven temperatures on the inner surface of the kiln, a failing incompatible with good-quality ceramic surfaces. An outside adviser to the cooperative had, through simple experimentation, discovered a simple and workable electric oven. Nevertheless, this specific technology was rejected because it necessitated bringing to the village a portable electric generator which only the cooperative could afford and which only a very few people could fully understand, maintain, and repair. The principle invoked to justify the decision was that only those technologies are "appropriate" which are in harmony with ancient Quechua rural values of mutual help and sharing the benefits in all improvements. After lengthy deliberations, it was decided to adopt a kerosene-fueled oven and to experiment with ways of improving the refractory (or heat-insulating) properties of local clay. The reason behind the choice is that all villagers already possessed prior experience with kerosene, and even the poorest among them could afford the kerosene oven.

The second principle which departed from conventional norms practiced in cooperatives affects the distribution of net surplus earnings. Here again, so as not to create social and economic distance between the producing cooperative and the larger village community, it was decided to assign a share of the surplus to all members of the village, whether they belonged to the cooperative or not.

Both principles have been applied in the wool cooperative as well as that dedicated to ceramics. Interestingly enough, the peasant associations receive partial outside funding.[61] Moreover, the local cooperatives are fully aware of their need to receive limited "technology transfers" from the outside. Nevertheless, for reasons pertaining to the revitalization of their cultural values, they have established a practical criterion for exercising control over the entry of outside technology into their community in ways which harness it to their self-perceived broader value goals. The operation is admittedly small in scale and has not yet proven its viability over long periods of time. Thus far, nonetheless, it clearly illustrates an important principle expounded in a theoretical vein elsewhere in this book: namely, the existence of a vital nexus among value options, development strategies, and concrete policies for the acquisition and assimilation of technology. These Quechua communities in Bolivia have deliberately and explicitly chosen to subordinate technological efficiency to their wider and more basic cultural needs. They have translated ancient Quechua ideals of solidarity and mutual benefit into a working instrument to guide decisions of a financial and technological nature. *Mutatis mutandis*, it is precisely this kind of approach which is required even of policy-makers in macrodecisional arenas. However modest in scope, the Tiataco-Huayculi experiment is qualitatively important and has value to others as a paradigm.

E. An Experiment in Transferring Technology within a "Developed" Country

Several approaches tried within the United States to transfer technology from one sector of activity to another shed light on constraints met in less-developed countries. Especially interesting is the technology transfer program conducted by the city government of Tacoma, Washington, and known as Totem One. This project, funded by the National Science Foundation, the Bureau of Standards, and private business, aims at enabling a municipal government to institutionalize the transfer of technological innovations made in the aerospace industry to such municipal operations as firefighting, court-scheduling, personnel management, development-planning, information systems, and law enforcement. Dual emphasis is placed on adapting hardware and developing new operating procedures.

The project is described in publications issued by the office of Tacoma's technology coordinator.[62] A few of the principles which guide the Totem One program are worthy of attention. According to joint evaluators, the best technique for achieving technology transfer from the Boeing company to the City of Tacoma is the "process approach." City personnel and aerospace technologists work together to develop mutual confidence. Out of such daily contact come projects and applications which are simultaneously important to the city and lie within the company's technological capabilities. The city has learned that it is futile to have technology salesmen look at its needs; what is required is daily proximity and collaboration between technologists from the transferor company and officials from the transferee city government. Most important, the private company must share the financial risk of shaping technological adaptations which can be used by the city. The city will not purchase new technology unless the supplier has successfully harnessed the pre-existing technology to some city operation, with clear indications that money will be saved or efficiently increased.

A wide consensus now exists that technology developed by private industry in the United States is not being optimally used outside industry. Hence, financial support from the federal government or private foundations is needed, in most cases, to subsidize technology transfer to cities. The number of cities which are receiving such support and attempting to replicate, at least in part, the Tacoma experiment, is growing rapidly. Thus the city itself comes to be viewed as an urban laboratory. The lesson is that, even within the United States, technology transfers do not function simply on commercial market lines; government subsidies and deliberate policy intervention are required. All the more reason why promoters of transfer in less-developed countries should recognize the role of deliberate science- and technology-planning allied to subsidies operating outside pure market mechanisms.

* * *

Technology policies are discussed in later chapters. Before they are, however, some attempt must be made to assess the price paid in social dislocation and human suffering by "receiving nations" for their technology transfers. This assessment, however tentative, must take into account the constraints at work in the mechanisms and channels for technology transfer from industrial to Third World countries. These mechanisms have now been examined, as have the criteria employed by transnational corporations as suppliers of technology. And the case studies concretely illustrate the workings of these mechanisms and criteria. The high price paid by Third World societies for technology transfers is the topic of the next chapter.

6/The High Price of Technology Transfers

Technology imports severely tax the hard-currency reserves of poor countries. A recent United Nations report

> places the direct cost, consisting of payments for the right to use patents, licenses, process know-how and trademarks, and for technical services needed at all levels from the pre-investment phase to the full operation of the enterprise, at about $1.5 billion in 1968, and further calculates the cost to be growing at a rate of about 20 percent a year.[1]

This estimate is probably too conservative. An editorial in the Brazilian newspaper *A Voz do Brasil* dated 3 January 1975 states that in 1974 Brazil alone spent more than one-half billion dollars for the acquisition of product technology in the form of equipment and machinery. Therefore, most Third World nations seek to import technology at lower costs. Many measures adopted by Andean Pact countries are explicitly designed to lower these costs.[2] Financial costs of technology are not, however, the central issue; more important are human and social costs of technology transfers.

The aim of this chapter is not to measure but to call attention to these costs, for technology transfers are often discussed as though they did not exact heavy social sacrifices. Even if they cannot be eliminated altogether, these costs must be carefully weighed when decisions are made. Special attention is given to the following considerations in evaluating social costs of technology transfers: their degree of compatibility with development goals, their impact on the quest for greater autonomy, conflicts over equity and social justice, the creation of jobs, and considerations of ecology and demography.

Compatibility with Development Goals

Technology Transfers: Aids or Impediments to Achieving Basic Development Goals?

The relative priority of goals any society pursues in development is central. Although general statements of goals are found in development plans, in practice planners make preferential choices indicating to which categories of people *greater material welfare*—a typical general goal—is likely to accrue. For instance, if a policy of favoring heavy industry is adopted to the relative neglect of agricultural sectors, it may be that importing "high" technology from transnational corporations is a channel of transfer fully compatible with this goal. If, conversely, the priority development goal in Country X is to equalize incomes among social classes, importing technology via ordinary channels may prove contrary to the desired objective. The difficulty is compounded because national development plans often hide budgetary priorities behind generalities about higher material standards and justice for all whereas, in fact, they favor limited sectors of the population.

Another broad goal usually sought by development planners and politicians is to endow their country with *"modern" infrastructure*: modern schools, an efficient public bureaucracy, statistical services and a tax administration, good roads, electric power, potable water, and communications systems. But countries vary in their preferences as to the degree of concentration of infrastructure investment. If the choice is made to decentralize widely, to create secondary and tertiary urban poles of development and offset the exaggerated pull effect of the primary poles, then conventional technology transfers will probably conflict with this goal.[3] The reason is that TNCs prefer to invest where modern infrastructures already exist, and they are enthusiastic about the "small" markets found in many underdeveloped areas. Hence their reluctance to invest in secondary or tertiary poles. Even when their role is simply to license technology to LDC clients, large TNCs favor large-scale national partners. Thus do prevailing modes of technology transfer place obstacles in the way of decentralized investment policies by LDC governments. Because possibilities of conflict abound, those who negotiate the acquisition of foreign technology need to examine the impact of their acquisition on efforts to decentralize infrastructures, especially those supportive of industrial activity.

A third developmental objective often explicitly or implicitly invoked by planners is the *transformation of values* among their populace. Literacy campaigns, the educational system, and general dissemination of certain images of the good life (smaller families, more spacious homes) are designed to change people's aspirations, values, and behavior. Yet influential decision-makers often fail to assess the "coefficient of impingement on values" of their projects or campaigns. Value options of "modernizers" within a country, and the popular reaction to their choices, will determine whether the technological values imported along with products, processes, and expertise will harmonize or not with development values sought.

A fourth development objective often professed is *self-sustained growth*. But is continual importation of industrial technology from transnational corporations consonant with this goal? Two issues need to be examined: (a) the shift in the locus of balance-of-payment difficulties from import substitution to payments for technology and (b) the bias towards a certain kind of growth inherent in competitive corporate technologies. How "self-sustained" can growth be if it depends on massive imports of technology for its dynamism? How long must technology be imported before it can be produced locally? ("Produced locally" means produced under national control for national purposes, not merely produced at local installations which remain under the control of outside firms. Steps for creating research and development facilities in underdeveloped locales are proposed in later pages.) The pertinent point is that a high price is paid for imported technology in part because the factors which make for continued dependency are considerably reinforced by continual reliance on outside suppliers of technology. Robert Girling correctly states that "the transfer of technology has proved to be a subtle and pervasive mechanism in the preservation of structures of dependency in the Third World."[4] But such an effect is not intrinsic to technology per se: it merely ensues from present commercial modes of technology *transfers*. Technologies sold by TNCs favor growth with huge scale, high concentration, and built-in obsolescence. Each of these features may prove to be antidevelopmental and inimical to the demands of distributive justice.

The *creation of jobs* is yet another widely professed development objective. Yet is is doubtful that modern technology can contribute to increasing employment.[5] José Walter Bautista Vidal, Secretary for Technology in Brazil's Ministry of Industry and Commerce, has declared that his government does not expect to be able to reduce unemployment in the primary or secondary sectors but mainly in the tertiary (services) sector of the economy.[6] Therefore, he concluded, it is futile to advocate labor-intensive technologies for industry. To adopt such technologies, he added, would simply render Brazil noncompetitive in world markets. At least Mr. Vidal recognizes a possible conflict between job-creation as a development objective and current technology transfer practices oriented away from tertiary sectors, so that, in the Brazilian case, one cannot point to an inconsistency between the country's overt employment policy and its approach to technology. Any criticism, if warranted at all, must be directed to both. One remains skeptical, however, as to the capacity of the secondary (manufacturing) sector to create new jobs at the rate of 5% yearly, as targeted in Brazil's Second National Development Plan (1975-79).[7] Brazil's recent performance suggests that employment policy is adversely or favorably affected by the mode in which technology is acquired from abroad.

The *economic integration of disparate sectors and regions* within a country often constitutes an important objective of national development strategies. If, however, no energies are spent to create "appropriate" technology with a high incidence of sectoral or regional factor composition, this goal turns out to be more rhetorical than real, especially if agricultural technologies need to be locale-specific.[8] Overstandardization is both economically disastrous and culturally destructive. Therefore, if overall costs of technology are calculated, outlying regions or weaker sectors of the economy cannot be sacrificed. Unfortunately, even within the agricultural sector, conventional biases favoring the large scale readily win out. But large agribusiness technologies may hinder the "modernization" of areas better suited —economically and socially—to smaller-scale farming. Thus habitual modes of technology acquisition from abroad conflict with still another developmental objective: *bringing economic dynamism to poorer agricultural regions.*

Development planners usually advocate *industrialization* and *capital accumulation*. A few countries, such as India, Brazil, and Iran, aspire *to become major actors in geopolitical arenas*, an ambition which propels them into seeking certain categories of technology over others. India, for example, wants an autonomous nuclear capacity; Brazil, the infrastructure needed to produce sophisticated weapons. But the channels of technology acquisition which lead to influential status as a global military or political actor favor the great powers which already hold an overwhelming advantage in the sale of arms.

A few Third World countries include *wider political participation* of their people among their development objectives. The Iranian government recently signed an agreement with Stanford University calling for the installation of locally orbiting satellites to provide instant television and telephone communications to 20,000 villages.[9] These facilities, it is claimed, are to be used for educational purposes, the rationale being that communication with events in the outside world is an important means for rapidly "modernizing" adults and children in remote hamlets. Yet the perilous ease with which such facilities can be used for political surveillance casts a dark shadow on the project and raises moral doubts as to its advisability. In this, as in myriad other cases, technology transfer exacts a high price.

The list of development objectives affected by the manner in which foreign technology is imported can be expanded to include:

- becoming a more efficient producer
- reducing dependency on the outside
- eliminating absolute poverty and relative deprivation of poor masses
- achieving greater social justice (expressed as proportionate shares of total assets, income, gross national product, social services, etcetera)

One general conclusion emerges unmistakably: *Sacrifices in reaching these goals are usually part of the price paid for importing technologies from the rich world.*

Modes of Impediment: Direct and Indirect

Furthermore, if uncritically conducted, technology transfers can actually *impede* the achievement of stated development goals in numerous ways. At times the acquisition of foreign technology runs directly counter to the stated objective, as when a decentralization or an "intermediate poles" policy is sought. In these cases success is sabotaged by importing technologies designed for large-scale, centralized operations which are wasteful, expensive and ineffective in decentralized sites. Under other circumstances prevailing modes of technology acquisition frustrate development goals indirectly by pre-empting a disproportionate share of scarce public infrastructure funds. This is the case in Brazil.[10] There, in order to subsidize potentially efficient (read: "competitive on the international market") large industry, the government has invested large sums in providing infrastructure in transportation, communication, tax privileges, import credit, and site facilities. These sums are thus removed from alternative uses more congenial to other categories of productive activity. This is the complaint voiced by promoters of rural development as well—they are cheated, if only by default, of their due share of public investment.

A third way in which standard modes of technology transfer interfere with development objectives is by building into the change process the exaction of too high a social cost for the achievement of certain objectives. The Brazilian case is once again illustrative. No one can deny the country's spectacular aggregate economic growth in the last decade. (The World Bank sets Brazil's average annual growth rate [in GNP per capita] for the years 1965-72 at 5.6%.[11] With population growth occurring at an annual rate of 3.2%, the gross rate of growth peaks at approximately 8.8%, high by any standards.) Nevertheless, the elections of 15 November 1974 and subsequent political events have revealed a widespread feeling in Brazil that the price paid for such growth is far too high: political censorship and repression; "selling out to TNCs"; the neglect of the poor North East and the agricultural sectors, generally to the advantage of the already rich Center-South industrial areas (around the so-called "industrial triangle" comprising the cities of Rio de Janeiro, São Paulo, and Belo Horizonte); and the placement of the major burden of the growth on the lower classes. No simple comparative yardstick exists for deciding when the human price paid for economic growth is too high, and one can easily fall into special pleading when evaluating the respective "human costs" of competing development strategies.[12] If, however, those who are called upon to pay the price repudiate it, then that price is clearly too high. To pursue social benefits at *any* price is to assure that these benefits will cease to be beneficial: there are cutoff points at which the price paid is too high.

In addition to the general costs attaching to technology imports just outlined, there are specific arenas of overt conflict which must now be considered.

Technology Transfer and Social Justice

Achieving greater equality or equity for the entire population of a country is rarely a priority goal set by development planners. It is praised rhetorically, but in practice, as judged especially from budget allotments, it is unimportant in developers' minds. Of course, there exists no single, or simple, yardstick for measuring "just" development, but in general terms true social justice embraces at least three elements: *equality, equity*, and *participation*. A development strategy which stresses social justice, then, seeks to achieve relative equality in the provision of basic goods and opportunities, is concerned with a fair distribution of the fruits of progress, and institutionalizes the concept that respect must be shown by leaders for the wishes of the people at large. Without my analyzing in theoretical fashion the requirements or the epistemological foundations of societal justice (others have done this well[13]), we surely understand that one is entitled to ask: Who benefits from technology transfers from TNCs to firms, laboratories, universities, and governmental agencies in less-developed countries—a handful of privileged professionals or large numbers of the populace? Unfortunately, there lie ready at hand no statistics analogous to those cited by Robert McNamara when he launched the World Bank on the course of attacking poverty in the lowest 40% of the Third World's population.[14] Little empirical knowledge is available to help us determine who truly benefits from technology transfers; we are forced to rely heavily on an analysis of structural trends. Müller cites a study by Adelman and Morris showing that greater inequality of income distribution and increasing concentration of wealth in the hands of the privileged usually occur in the first years of economic development.[15] Yet we are given no information which would help us trace this process of inequalization directly to the technology-transfer phenomenon.[16] Nonetheless, it seems obvious that if technology transfers do not benefit the masses, the reason is that they are not *designed* to benefit them, but rather to create marketable new products and processes. By definition, of course, a "market" is where effective purchasing power lies—in the hands of consumers whose basic needs are already met. Modern technologies are, indeed, best at producing "marketable" goods and services so expensively priced that they are out of reach of those most in need. It only follows, then, that transference of these technologies impedes social justice by contributing nothing to enhance it and, furthermore, by siphoning off resources for lesser priorities. Technology transfers, as now conducted, tend only to improve the relative position of those who "benefit" directly from them. And by bettering the relative position of those already favored, they worsen inequality. But inequality exists at many levels: assets, income, consumption, and opportunity.

In terms of assets, technology benefits primarily individuals and institutions already in control of large amounts of resources. Expensive technologies cannot be afforded even by poor firms, let alone by individuals. Most products and services facilitated by technology enter into the "basket of consumer goods," to use Celso Furtado's phrase, which can only be purchased by the rich sectors of poor societies.[17] As for levels of income, technology transfers obviously reward engineers, chemists, and technicians more generously than the unskilled and, *a fortiori*, the unemployed. At the level of opportunity, the issue is linked to the overall educational and training systems entrenched in given societies; that is, unless these systems are explicitly restructured with a view to making them equitable in less-developed countries, technological opportunities in most cases will be monopolized by the tiny apex of the educational pyramid.

As one turns to degrees of "participation" in technological innovation, design, and operation, it stands out clearly that modern "imported" technologies exclude, by definition, unskilled workers. Nevertheless, most plant managers and personnel officers interviewed in Latin America declared that the mere introduction from outside of a new technology—be it a machine or some piece of equipment—arouses the curiosity of many workers who want to learn enough to work with it. A certain fascination attracts workers not otherwise inclined toward routine technology and serves as an informal vehicle for intensive training which can quickly lead to "cultural accumulation," that is, a familiarity (diffused throughout the general work force) with machines, electricity, and chemical processes upon which engineers and other specialists depend.

No discussion of imported technology's impact on social justice, however, can avoid the issue of employment.

Technology and Jobs

The gap between policy rhetoric and reality in job-creation is revealed tragicomically in the following true story told by E.F. Schumacher, now acknowledged as the "father of intermediate technology":

> I was in a developing country not so long ago and was shown around a textile factory—the manager was a European, a very courteous man, and he said he was proud to show me this fac-

tory because it was one of the most modern in the world. I said, "Before you go on, can you tell me what's happening outside, because as I came through here there were armed guards there, and you are beleaguered by hundreds and hundreds of Africans." "Oh," he said, "take no notice of that. These are unemployed chaps and they hope that I might sack somebody and give them the job." I said, "Well, as you were saying, you have one of the most modern factories in the world." "Oh yes," he said, "you couldn't find anything better." "How many people do you employ?" "Five hundred. But it's not running perfectly yet; I am going to get it down to three hundred and fifty." I said, "So there's no hope for those chaps outside?" He replied, "The people demand perfect products and these machines don't make mistakes. My job is to eliminate the human factor." I then asked, "If you make such a perfect product, why are you here in this wretched provincial town and not in the capital city?" He said, "It was that stupid government that forced me to come here." I said, "I wonder why?" He replied, "Because of the unemployment in the provinces."[18]

Not all modern managers are as callous as the third-person "hero" of this tale, nor are all stewards of machines enjoined to "eliminate the human factor." Yet no greater source of friction pits less-developed countries against suppliers of technology than the issue of job-creation: this is the omnipresent abrasive. Arguments are often phrased abstractly around a question such as: Are modern technologies too capital-intensive instead of labor-intensive? Or: Do such technologies make optimum use of the abundant local-production factors (especially in poor countries saddled with excessive labor power)? Almost always the answer is negative. But a few refinements need to be introduced in the discussion.

Louis Wells believes that many less-developed countries could increase job-creation by using machinery which is not brand-new but more labor-intensive than up-to-date models. Such benefits, he thinks, could be obtained without undue sacrifices in efficiency. His research in Indonesia has led him to several conclusions of interest.[19] The choice of technology, Wells asserts, has direct implications for employment; he claims that, in certain industries, labor-intensive techniques can provide "more than ten times as many jobs as the capital-intensive plants for the same output."[20] But the choice of technology is not always dictated by a desire to keep costs at a minimum. Among other considerations which intervene is the easier access to credit enjoyed by foreign firms. No doubt it is dangerous to generalize prematurely inasmuch as some foreign plants display a greater tendency to use intermediate technologies than do certain locally-owned counterparts. More interestingly, "the need to produce high-quality output also [does] not appear to explain the differences in plant design. In most industries, high-quality products [are] produced in intermediate-technology plants as well as in capital-intensive ones."[21] Furthermore, the scale of operations does not appear to be related to capital intensity. The most significant variable affecting the choice of technology seems to be the competitive position of the firm. Where brand image is the basis of competition, a plant tends to be relatively capital-intensive. When, on the other hand, price is the basis of competition, pressures to reduce costs drives firms to a more labor-intensive technology. Plant designs are influenced by the desires of managers not to have to handle large, and unpredictable, labor forces, coupled with the bias shared by most engineers for sophisticated equipment.[22] When pressure from competition is weak, these managerial and engineering criteria play a dominant role in the choice of technology. The obvious lesson is that the selection of technology by firm managers is dictated by many criteria, some of which have nothing to do with the minimization of costs or the creation of jobs. A government can, of course, intervene to influence the choice of technology in desired directions. Nevertheless, even in projects run by government agencies, the criteria effectively invoked are surprisingly "noneconomic" or "nondevelopmental."

An interesting study by Harvard professor John W. Thomas on technological alternatives available to the government of Bangladesh in the late 1960s for implanting irrigation tubewells illustrates the principle. External financial assistance was available and, although detailed cost-calculations of several types of tubewells were made—incorporating such variables as drilling technique, power source, type of engine, type of pump, screen material utilized, and the drilling agent—it was finally decided to use the capital-intensive, less-than-optimal well. Thomas explains why:

> On balance the arguments for the low-cost wells over medium- and high-cost, appear impressive. With low-cost wells, economic return is higher, the employment and training effects are greater, the components of the wells hold greater potential for the creation of domestic industry and they will provide a broader distribution of the benefits of well-irrigation. This evidence, plus the fact that low-cost wells were the only ones proven in actual operation in East Pakistan suggests that the low-cost wells with percussion (or jet) drilling, brass strainers, centrifugal pumps, and low-speed diesel engine represented the logical tubewell technology for the country. The fact that the Government requested assistance primarily for medium- and high-cost wells and the aid donors almost exclusively preferred the medium-cost wells suggests that standards other than those examined are paramount in the decisions of Governments and aid-givers as the appropriate technologies for developing countries.[23]

What tipped the scales was not economic optimality—or even conformity with government policy—but the "organizational requirements of the implementing agencies, including the aid-donors."[24] And so a merely satisfactory, instead of the best available, solution is accepted because it is familiar to the agencies involved and because it minimizes risk.

Thomas's conclusion sheds light on procedures observed by most "receivers" of imported technology in underdeveloped lands. Whether they be private firms or public agencies, their desire for organizational control, their compulsion to minimize risks of failure, and their infatuation with "modern" technology all weigh heavily in favor of *not* adopting technologies which recommend themselves on grounds of job-creation. But in the final instance, the decision relative to labor intensivity is not itself technological: In most cases where choice is possible, outside criteria are what will influence the choice of a technology which ultimately "proves" to be job-creating or job-minimizing. This does not mean that any given technology is, in employment terms, indifferent. But it does mean that nontechnological values must be asserted and inserted into the decision-making process if job-creating technologies (assuming that these exist and are rationally defensible) are to be chosen. Required is an unflinching commitment by planners to attack the unemployment problem directly with those technologies best suited to do so. Wells believes that such a policy could make wider use of second-hand machinery, obtained mainly from other less-developed countries.[25] He found the origin of machinery to be closely related to the capital intensity of industrial processes (machinery reflecting the factor endowments of the country from which it comes). Consequently, countries like Indonesia, situated far from second-hand machinery markets in the United States or Europe, need *information links* to alternative sources such as Singapore, Taiwan, Hong Kong, and the Philippines. Here lies the key to using more machinery from developing countries. Governments themselves can offer financial incentives to firms which purchase labor-intensive equipment in developing countries. Duties, taxes, and credits can all help create or offset competitive advantages. Success will be limited, however, unless labor-creation and the selection of suitable technology are related to broader issues of employment structures.

This link between jobs and broader social processes is the theme of a seminal essay by Friedmann and Sullivan.[26] Their study is limited to labor-absorption in urban settings, but cities are the locus of most manufacturing industry and the chief importers of foreign technology other than agricultural and military technologies. Consequently, their analysis is germane to overall technology policies. Friedmann and Sullivan divide urban employment structures into categories each having diverse capital requirements, production and productivity scales, ease of access, income-generation potential, and labor-absorption capacity (measured in percentage of labor force). The importance of this classification lies in the policy measures suggested by a breakdown of the urban labor force into: *unemployed* workers; those employed in a city's *"street economy"* (the individual-enterprise sector, including a wide gamut of self-employed persons: handicraft workers; street traders, vendors, and service workers; casual construction workers; persons engaged in underground occupations—prostitutes, professional beggars, police spies, dope peddlers, pickpockets, etcetera); those employed in the *family-enterprise* sector (workers in small trade and service establishments and industrial workshops having fewer than fifty employees and a low capital-to-labor ratio); and those working in the *corporate* sector (workers in larger corporate enterprises, large family establishments, government bureaucracy, universities, and liberal professions). One major difference among these sectors is that the actors in the first three mentioned are for the most part self-financed, whereas those employees within the corporate sector have direct access to government revenues, bank credit, or corporate profits. "Partly as a result, capital intensity (and therefore labor productivity) rises progressively with each step in the hierarchy."[27] A more important distinction, however, is that each of three self-financed employment sectors functions as a distinct subsystem of the urban economy—characterized by its own economic attributes, social relations, and ethical rules—while the corporate sector enjoys the legal protection of "the system" (labor legislation, social security, and the like) in a measure far outweighing that extended to the "lower" sectors. Income gains in the two lower sectors and the less-protected portions of the corporate sector are precisely those which are quickly dissipated among new arrivals to the city. The main analytical conclusion reached is that

> if the lower two thirds of an expanding urban population are getting progressively poorer, the upper one third and, more specifically, the less than 3 percent of the population who derive their incomes from the P/M-subsector (professional and managerial personnel), will be the principal beneficiaries of continued economic growth.[28]

Therefore, balanced regional development, which shifts the primacy away from urban areas toward decentralized poles, is essential if the employment problem, even in cities, is to be solved. Only in such a context can efforts such as those made since 1969 by the International Labour Office direct policy beyond palliative treatment of symptoms and offer hope of reducing unemployment.[29] Robert Theobald, author of *The Guaranteed Income*, goes still further and deems even concerted action to be futile.[30] According to him,

> One of the few things that is perfectly clear about every developing country is that "full employment" is an impossibility. The only reason we ever reached full employment in the countries

> which are now developed is because technology required using all the people who were coming into the cities. Today, on the other hand, technology is so advanced that even if industrialization takes place it absorbs very few workers. Yet we are still trying to get full employment instead of accepting that today our only hope is to break the links between income and employment, to recognize that we must treat the problems of production and the problems of distribution of resources as separate problems.[31]

It is useless, Theobald argues, to try to create jobs, because the only way to create enough jobs is to accept a general level of productivity in the economy which will not produce enough basic goods for all. Therefore, he concludes, let societies—particularly developing nations—adopt the most modern technologies in order to produce as much as possible, but let them also simultaneously provide some form of guaranteed income to all their members. Theobald insists that the "green revolution," based on abundant use of fertilizers and pesticides, cannot solve the problem of increasing food supply for the poor. He prefers using "very high-level technology involving nuclear reactors, desalinisation where it is necessary, and chemical greenhouses to create a resource where there is none at the moment, rather than trying to restructure land use which is a process which has inevitably torn cultures apart."[32] He further asserts that the best technologies, properly employed and harnessed to the maximizing of human potential in a variety of cultures, can lead to the creation of "a society of enoughness in which people will accept that too much is just as destructive as too little."[33] His prescription, in short, is to attack the unemployment problem by ceasing to treat it as the *major* problem to be solved. His preferred solution is to maximize production and so structure distribution that the basic needs of all are met independently of their desire or ability to work at a paid job. This strategy has never been seriously tried in any national society, but it does introduce into all policy-thinking critical elements as to the relationship between technology and job-creation.

Paradoxically, Theobald's "radical" view rejoins the more classical position expounded by Oxford economist Frances Stewart when she reminds us that "the most labor-using technique in the short term may generate less employment in the long run than alternative techniques. Future employment depends on future levels of investment, as well as the capital intensity of techniques."[34] But Theobald pushes still further by arguing that short-term concentration on labor-saving technologies can produce neither the desired long-term employment nor the requisite productivity to meet the basic needs of all.

Professor Stewart raises an interesting point regarding the total costs of labor-intensive versus capital-intensive techniques. After noting that working capital is an important aspect of capital costs usually overlooked in discussions of the question, and that working-capital requirements differ widely according to mode of operation, scale, and labor requirements, she concludes that these costs "are likely to be proportionately heaviest for the most labor-intensive techniques."[35] If this is indeed the case, then labor-intensive technologies may turn out to be also capital-intensive. Other widely neglected variables, she adds, are the differences between urban and rural choices of comparable technologies and the importance of income distribution in the locality of a production site as affecting choice of technique with a view to product differentiation designed to meet that market. These elements pale into insignificance, however, in the face of a larger factor, namely, that "the transfer of advanced-country technology is in large part responsible for the growing employment problem."[36] Such technology has not only limited the possibility of job-expansion but has also helped create a demand for more jobs in two ways: by accelerating population growth (medical technology) and by fostering an explosion in aspirations (media technology). For this reason, the structure of any LDC economy should render appropriate (that is, labor-creating) technology profitable. Like other students of the question, Stewart concludes that job-creation must be directly attacked by overall development strategy, of which direct actions on technology are but a part. Unlike Theobald, she accepts the possibility of increasing employment in various countries by allying sound technology policies with restructured incentive systems and opportunity-providing institutions.

What conclusion emerges from research on the relation between technology transfer and employment? The answer is that current patterns of transfer exact a very high price in unemployment and underemployment in most underdeveloped countries. What seems no less clear is that this price cannot be lowered to tolerable levels, at least in populous less-developed countries, by altering the technology system in isolation from larger social transformations. These larger changes bear on overall incentive systems, research and educational structures, tax policy, political decisions regarding the type of production to be subsidized and otherwise supported, and strategies for locating productive investment in optimal patterns. Employment is a choice problem arena wherein the vital nexus between society's basic options, development strategy, and specific policy is most visible. All three condition not only the price in unemployment which will be paid but also who will pay that price—those least able to afford it, or others. At present only the affluent can afford the luxury of being unemployed.

Other Costs Incurred in Technology Transfers

Thus far I have focused on the high price technology transfers impose on poor societies in three domains—development objectives, social justice, and employment. There are certainly other social costs, however, and significant among them is the sacrifice in cultural autonomy that results from the tendency of modern technology to standardize products, processes, aspirations, facilities, work styles, instruments, and overall modes of living.

Cultural autonomy is difficult to maintain in conditions of rapid change, particularly among subnational groups. Indeed one of the most conspicuous social effects of the dissemination of Western technology is the homogenization of lifestyles. Standardization is evident not only in airports, hotels, and tourist installations but also in industrial parks, residential suburbs, supermarkets, clothing, habits of food consumption, aspirational levels of professionals, and other domains. Theorists of leisure like Veblen, Sombart, Pieper, and Huizinga long ago called attention to the role-playing "emulation" of the rich by the poorer classes. But conventional technology transfers not only recondition psyches; they also alter the basket of consumer goods concretely available to large numbers of people. We have read, for instance, about an American consultant firm's preliminary plans for satisfying (if not creating?) a mass market for US-style TV dinners in a Latin American country.[37] What an anomaly in the context of the unimpeachable rhetoric of the company's document, which speaks of the widespread wasteage of food resulting from poor refrigeration or bad storage facilities and the consequent needs to assure that produce gets to the table of the underfed. But the lure of standardization is powerfully inscribed in the very logic of multicorporate technology. The trick consists of taking a genuine human need, packaging it in some manner advantageous to the supplier, and capturing the aspirations of the population at large so that its generic desire will be expressed as a compelling urge to buy the specific package. This is exactly what large corporations do in order to create markets for new products as well as for old products which are only slightly transformed but are made to appear radically new and different. Therefore, the psychological energies unleashed by the experience of thirst are pre-empted by Coca-Cola; the need for transportation, by manufacturers of automobiles; the dream of a vacation, by travel agents who convince people of their "need" to fly far away via expensive airlines, and so on. Thanks to its influence on the aspirational content and schedules of large masses of actual and potential purchasers, modern technology (particularly advertising technology) deeply affects popular cultures in most less-developed lands.[38]

The culture of any society expresses itself in the modes of work and of relating work to leisure it adopts. Modern technology, once transferred to matrices other than those of origin, imposes its logic of uniformity on tools, work paces, and safety standards. In one of his early works, *Technics and Civilization*, Lewis Mumford attributes a special influence in this direction to the technique *par excellence*, the clock.[39] Yet the elements related to work and leisure are far more constitutive of a culture than many of the externals outside visitors emphasize: picturesque dress, exotic music or festivals, colorful wares artistically displayed at itinerant fairs. (As one member of the British Parliament puts it, "Culture, after all, is about people and patterns of everyday life—not monuments and souvenirs."[40]) Some procedural uniformities are doubtless inevitable; but development planners, as they make their choices, should beware of the high price on cultural destruction exacted by modern technology. If they are truly concerned with preserving cultural diversity, they must select machinery and other work-related technologies which protect diversity. Their decisions have great bearing not only on the quality of work and its meaning in people's lives[41] but also on their patterns of consumption, the degree of urbanization deemed acceptable in their societies,[42] and the scale of the institutions they will choose. These are the vital *loci* where cultural survival will be assured or lost (which is to suggest not that the fine arts are inconsequential but simply that they are easily relegated to the periphery of cultural values, or themselves altered, when technology sets the pace in daily living). Nowhere do the values vectored by modern technology so quickly assert their primacy, or win adepts, as in the behavior of business and professional elites. Not only their language but their dress, ethical codes, and stylistic preferences rapidly become modeled on those of rich-world counterparts. Manifestly, this standardization is not always or necessarily to be regretted. Yet, if one accepts the view that such elites increasingly constitute the sociologically "significant others" to which masses refer in their aspirations, one is less than sanguine about the viability, over the long run, of a plurality of rich cultures.[43] One may continue to meet in La Paz, Nairobi, or Teheran peasant women in traditional garb alongside bankers in ties and business suits. Although such picturesque residual symptoms of cultural diversity may long coexist, the real question is: Whose values are dominant in the elaboration of school curricula or the programming of radio and television? Will the children of the Bolivian women be more powerfully influenced by the engineer's values and culture than the engineer's children by her Quechua values? The answer programmed by most societies is easy to give, all the more so because most education ministries in poor countries have themselves joined the race to harness technology to teaching in their schools.

Indeed technology transfers impose a very high price in cultural dependency, a price which can be minimized by deliberate policy measures only if cultural homogenization is recognized as a serious danger inherent in uncritical technology transfers.

Another realm wherein transferred technology exacts a price from nonindustrialized societies is that of ecological integrity. Since the 1972 United Nations Conference on the Human Environment in Stockholm, a veritable flood of documents on ecology and development has come forth. In some we are told that the Third World has a right to pollute in its race to industrialize. Conversely, we are informed by others that rich-country corporations are irresponsibly "exporting pollution" to the Third World and that the protection of the environment is a luxury poor countries cannot afford. The gap in basic perceptions of rich and poor nations is laid bare in the list of assumptions adopted by the Bariloche Foundation (Argentina) in response to rich-country models of limits to growth.[44] Bariloche's assumptions are that:

(1) The catastrophe predicted by the MIT Meadows model (hunger, illiteracy, poor shelter, etcetera) is an everyday reality for a great part of mankind.

(2) A policy of preserving the ecosystem is not possible until every human being has reached an acceptable level of life.

(3) Human development is blocked not by material limits to growth but by sociopolitical distortions in power distribution among classes and nations.

(4) It is neither possible nor desirable for poor nations to follow the same road as that taken by today's "developed" societies, which have engaged in wasteful consumption, accelerated social deterioration, and caused increasing alienation.

(5) A reversal of deterioration in the ecosystem can come not from mere correctives but only from the creation of a society intrinsically compatible with its environment.[45]

The issue is one of social justice. Who should pay to preserve the ecosystem? Should the price be borne mainly by those who, by their wasteful growth in the past, have depleted resources and continue to consume them voraciously, or by those who have only recently begun to use depletable resources and pollute the environment as they attempt to grow? Generally speaking, representatives of rich countries and international agencies plead for a global bargain to assure patterns of resource use which protect spaceship earth from ecological catastrophe. Unfortunately, their preferred scenarios do not call for a rapid and concerted attack upon the poverty of the world's masses as prelude or accompaniment to the protective measures they advocate. Therefore, their recommendations elicit rebuttals like those outlined in the Bariloche statement. Increasingly, Third World statements[46] on ecology highlight the following principles: major costs of protecting the environment, the resources, and the viability of the planet ought to be borne by those rich countries which have most egregiously depleted the earth's goods; any new global strategy for resource-use ought to acknowledge that poor countries need to increase their output rapidly so as to abolish mass misery; no informal "club" of rich countries should dictate which growth rates are appropriate for any LDC. Third World leaders neither deny nor ignore the gravity of the ecological problem, but they do resist rich-world diagnoses and prescriptions.

Although international cooperation is not precluded by Third World governments, they claim the sovereign right to define the problem in terms of their own developmental priorities. Moreover, they judge the position of rich countries on demography and ecology issues to be too aggressive and unilateral. The rich world's seeming obsession with these topics is viewed as a smokescreen behind which the privileged will continue to domesticate Third World development aspirations. An examination of policies in individual countries, however, displays bewildering variety: There exists no uniform approach by less-developed countries to ecological dangers, and some rich-world formulations of the problem have recently incorporated elements of Third World themes into their diagnoses.[47]

Many Third World countries want rapid industrialization and are relatively unconcerned about pollution or resource depletion. They welcome investors—domestic and foreign—even when these flaunt their disregard for ecological integrity. Ideological preferences affect positions taken by nations on ecology issues. Within the United States some critics contend that environmental irresponsibility is directly traceable to capitalists' disregard for social values in their quest for maximum profits.[48] The implicit assumption is that socialism is intrinsically more responsible toward the environment and larger social values. Careful distinctions need to be made, however, on this score.

The Soviet Union and China stand as paradigms of contrasting socialist approaches to ecological integrity. If Victor Ferkiss is right,

> the record of the Soviet Union and Eastern Europe in the field of environmental pollution is as bad as or worse than that of the capitalist industrial nations in part for technical reasons, since socialist accounting makes calculation of negative externalities even more difficult than it is under capitalism, but primarily *because growth has become the great socialist god.* The fetishism of commodities which Marx condemned as a feature of capitalism is just as strong a force in the socialist world as in the West.[49]

China, on the other hand, rejects mass consumerism, although it seeks growth in production and productivity. By deliberately subordinating growth to the inculcation of revolutionary consciousness and an emphasis on the primacy of moral over material incentives in economic effort, China is better placed to *internalize* ecological considerations in its calculus of social costs. As Orleans and Suttmeier write, China's insistence on social justice in conditions of great scarcity inevitably leads it to be protective of all its resources.[50] Therefore, it will necessarily recycle products rather than throw them away; it will minimize waste.

Within less-developed countries, one discerns a similar correlation between the option for austerity, understood as sufficiency for all, and the degree of tolerance one shows for ecologically damaging practices. Those countries which choose the capitalist road to industrialization and rapid aggregate growth argue that their initial levels of contamination are low and not too dangerous. Yet the problem has reached such proportions in such urban centers as Mexico City and São Paulo that one is skeptical about these claims.

The ecological argument lends itself to many uses, however. In Puerto Rico it served as a rallying point for diverse groups—partisans of independence, church organizations concerned with pollution and social justice, native industrialists eager to counter the power of foreign investors, and members of opposition parties in search of an issue to embarrass the government in power.[51] They merged forces to resist the opening of two large copper mines on the island by Kennecott and American Metals Climax, Inc. The fight later shifted ground to the desirability of building a superport for oil tankers on the west coast. Arguments linking ecological damage to social justice and the low general benefit to the poorer populace were employed in both cases against the respective claims, based on economic considerations, of the companies and the government.

No general conclusions may validly be drawn from this or similar cases as to the stance of Third World nations on ecological issues—air and water pollution, depletion of nonrenewable resources, disfiguration of the land, and the extinction of living species. Apparently most transnational corporations conduct their affairs in the Third World with far less regard for ecological health than they are obliged to show in their home countries, where they are constrained by more stringent legislation, better-organized public opinion, and the greater need to project a public image as "responsible" investors. Otherwise stated, foreign investments and technology transfers exact a high ecological price in the Third World.

The related issues of ecology and resource-use have grown more urgent and are more intensely debated in recent years because of increases in food, fertilizer, and fuel prices. Neglected arguments about alternative fuel uses take on a new topicality. Certain countries, it is true, are generously endowed with water power for their energy needs, while others have abundant petroleum reserves. But many others have to confront rapidly increasing fuel needs without abundant hydroelectric potential or thermal fuel deposits. Therefore, they too, like the rich industrialized countries, become interested in research on alternative-fuel technologies. Among alternatives, the use of solar energy for major percentages of one's fuel needs has recently received serious attention.[52] In its simplest terms, the argument is that fossil fuels will eventually run out. And because nuclear fission as a source of energy is both expensive and dangerous, solar energy and nuclear fusion are proposed as alternatives which are both less expensive, in the long term, and tantamount to inexhaustible. Alternative-fuel possibilities have recently been under study in a variety of underdeveloped countries, including at least some (Bolivia and Saudi Arabia) where petroleum is abundant. There is no reason to doubt, therefore, that before many years have passed suitable technologies rendering solar energy and nuclear fusion will become available. Accordingly, Third World policy-makers are well advised to assess the present price they pay in ecological sacrifices attendant upon fuel-related technology transfers in the light of these future possibilities.

Other domains where technology transfers impose a heavy price are medicine, contraception, military technology, and communications. As the purpose of this chapter is not to analyze each in detail, it suffices to mention them here. Nevertheless, the scale, quality, and cost of medical, contraceptive, military, and communications technologies have great bearing on who benefits from the services they allegedly provide, on what proportion of a government's funds are channeled to developmental purposes as distinct from weapons, and on the degree of educational and recreational autonomy any society can maintain. The task of measuring and comparing social costs incurred in technology acquisitions from abroad is no less complex in these domains than in those of industry or employment, although these costs are perhaps more visible in the latter realms.

* * *

This chapter has, in short, argued that technology transfers between rich and poor countries, as presently conducted, result in very heavy social and human prices in receiving societies. Most of these costs are not readily measurable and, often, not easily detected, but they are, nonetheless, real. More importantly, they are not all inevitable. One great merit of the "intermediate technology" movement lies in showing that these costs can be lowered. Because the costs are high, many assert that Third World societies should not uncritically receive technology from the "developed" world but strive to become capable of creating their own technologies in harmony with values they cherish. Once they begin doing this, they will have strengthened their capacity to receive even foreign technologies in a more creative and less destructive fashion. They may even assist rich countries to discover pathways to technological wisdom, for in truth technological development exacts heavy tribute from exporters, as well as from importers, of technology. Both need to find ways of lowering this exaction. Indeed to lower the price paid, in sacrificed values and human suffering, is one of the primary objectives of any technology policy.

Technology policies aimed at optimizing development benefits and lowering unnecessary human and social costs are most urgently required in Third World countries. Part Three, accordingly, inquires into the content and the context of suitable technology policies for development.

Part Three:

TECHNOLOGY POLICIES FOR DEVELOPMENT

Introduction

After exploring the value content of the technological universe and the ways in which technology circulates from rich to poor societies, this book now inquires into questions of technology policy for development.

This policy discussion is premised on the view that the vital nexus which links the value options of a society to its preferred development strategy, and to the criteria it adopts for problem-solving in specific policy areas such as technology, should be as explicit and coherent as possible. Although full coherence is rarely achieved by planners or those entrusted with implementation, it is worth striving for; technology policy will fail unless it is reasonably consonant with larger value options and change strategies. Technology policy embraces a vast network of domains relating to a nation's scientific and technical pool, material and financial infrastructure, overall incentive system, attitude toward outside agents, degree of control over the direction and speed of planned social change, level of integration into global or regional economies, and relative priorities attaching to technological modernity itself.

Indian economist S.L. Parmar warns poor countries that the international relations which draw them into the orbit of richer countries pose special problems. "While the benefits of their prosperity do not easily flow towards us," he writes, "the spill-over of their adversities tends to impose disproportionately heavy burdens on our economies."[1] This spill-over manifests itself acutely in technology transfers from rich to poor. In Parmar's catalogue of ills, technology aggravates unemployment, skews the distribution of income, increases dependence on outsiders, thwarts indigenous innovation, and favors counterdevelopmental trends such as high consumption and obsolescence in the design of goods. Transfers, in short, use resources wastefully in ways unsuited to sound development. Hence, many developing countries now recognize that these undersirable consequences abound because they lack a well-articulated technology policy which links development objectives to the dynamics of technological innovation and to prevailing patterns of transfer. Technological policy-makers in the Third World need to reflect critically on the principles underlying development strategies, that is, on the basic value options of their respective societies. These options, as they relate to development strategies, constitute the theme of Chapter Seven and prepare the way for outlining sound technology policies and practical modes of implementation (Chapter Eight).

National policies, however, are never framed in a contextual vacuum; they respond to myriad forces originating outside national borders. Hence Chapter Nine explores the impact of the changing international order on Third World technology policy. Claims and counterclaims vie for legitimacy in global forums where development issues are debated. A plethora of alternative models for a new international order is proposed. The purpose of this work is neither to review nor to analyze these alternatives but to examine the values of major actors in the international arena and to reflect on new forms of multiple loyalties required if changes in the international order are to foster sound development.

7/Development Strategies: Basic Options

The risk of oversimplification is great in any classification of development strategies theoretically available to planners. Much confusion results from debating at levels of generality which render comparisons meaningless, as when some specialists speak of an urban versus a rural strategy or one aimed at growth against one stressing equitable redistribution. By definition, however, development strategy comprises the totality of social changes amenable to planning and stimulation. Thus it matters greatly which basic images underlie one's diagnosis of underdevelopment and view of the development process. Marshall Wolfe, a social planner with the United Nations Economic Commission for Latin America, postulates three basic images of the development process, to each of which corresponds a preferred strategic path.[1]

A first image is that of a straggling procession of countries scurrying in vain to "close the poverty gap." Most weaker nations and most poorer masses within nations fall behind: their relative and often their absolute position worsens. They constitute, in the denigrating language used by some Western writers, the "soft" states which "can't make it."[2] Reasons adduced for probable failure are multiple: the "trickle-down" of development benefits takes many decades; policies to promote growth inevitably worsen the lot of all except the highly productive "modern" minority; international or national policies do not directly attack the poverty of the poorest or mass unemployment; growth itself is but a "modern" mask to perpetuate inequitable privilege systems. The image of the straggling procession suggests policies aimed at "closing the gap."

A second image, development as a living pyramid, lends itself either to conservative or to revolutionary interpretations. Countries, classes, and interest groups on top of the pyramid rise higher or keep their lofty position because they rest on the shoulders of majorities whom they exploit. As a living structure, the pyramid is in constant movement caused by the endless jostling of competing groups for position. This pyramidal image applies both to domestic class relations and to international distribution of wealth, power, and influence. Conservatives accept the image both as a portrait of what is and as a legitimate defense of what ought to be. Revolutionaries, in turn, while conceding that the pyramid exists, deny its legitimacy. They seek to alter drastically the configuration of the social structure so that no minority privilege group, old or new, can rise to the top. Their goal is to stratify society so that there will be no "top" where some privilege group can gain solid footing. The development strategy of each group flows largely from its own diagnosis, resulting in a decision either to engage in stable, incremental problem-solving or to subordinate all problem-solving to the radical alteration of power structures, respectively.

The third image in Wolfe's typology portrays the development process as an apocalypse: what Robert Heilbroner calls the Great Ascent is headed not toward the Promised Land of Development but toward the Bottomless Pit of Catastrophes—ecological, biological, psychological, and political. Subscribers to this view emphasize limits to the carrying capacity of the ecosystem; impute alarming dangers to rapid population growth; and fear global destruction through radioactivity, nuclear demolition, or the use by state agencies of biological technologies for purposes of social control in ways destructive to human freedom or even to human consciousness. This third imagery rejects two basic assumptions which the other two views, notwithstanding their great divergences, hold in common, namely, that long-term growth in production is good and desirable and that technology has an unlimited capacity to solve all problems—even those it helps create. Like the others, this third view prescribes strategies consonant with its diagnosis and posits criteria for deciding which developmental tasks are primordial. The stress is on renewing depleted resources, achieving zero population growth, and harnessing technology to a steady-state economy.

All efforts at diagnosis and prescription center on value judgments about what a good, or a better, human society is. For some a better society is one in which greater access to opportunity, if not to more tangible benefits, is created; for others the goal is effective equality in modes of greater or lesser participation; for still others the basic aim is to assure the planetary survival in modes which salvage human liberties. The three images are not mutually exclusive, nor are they always found in their pure state. Nonetheless, Wolfe's typology helps to focus our attention on classifications which transcend purely ideological or programmatic preferences. In 1972 Mahbub ul Haq, a World Bank economist from Pakistan, predicted that "the days of the mixed economy are numbered. The developing countries will have to become either more frankly capitalistic or more genuinely socialist."[3] Perhaps so, but no demonstrable correlation can be found either between the ideological system adopted by a poor country and its ability to close the gap or between the degree of class stratification tolerated internally and a country's degree of success in solving its ecological problems. Thus although the People's Republic of China has introduced great equality, socialist Algeria has not done so.[4] And class differences in the USSR may be quite as great as they are in Brazil, although they are certainly based on different social attributes.

More crucial than the ideology it espouses, however, is whether a society conceives of development merely as the pursuit of certain benefits or as the quest of these benefits *in a certain mode.* How benefits are obtained is as essential to defining development as the fact *that* they are obtained. This is not to gainsay the importance of benefits sought—greater material welfare, higher production and productivity, more efficient institutions, the growing ability to sustain dynamic economic performance. Yet it matters enormously *how* these gains are sought or obtained: in a pattern of high, or of low, dependency on outside powers; in a relatively equitable distributional mode or in ways which enhance the privileges of favored minorities to the detriment of needier masses; in a paternalistic, impositional style or in ways which progressively empower the populace to choose its targets and the instruments to reach them. Indeed, if true development is to take place, hitherto passive objects must become active subjects of change, and larger institutions must enhance their ability to participate in decisions affecting development. All these values refer to the *mode* of change, not primarily to its targeted *content*. The relevant point here is that varying images point to diverse policies indicating *how* developmental benefits will be sought. This is true even when agreement exists as to the desirability of the goals of effort.[5] Among central questions affecting the mode of development are these:

- Which institutional arrangements best promote development goals (politically centralized or decentralized, degree of coercion in planning, etcetera)?
- What relative roles are to be assigned to political leaders, experts, technicians, and "the people?" (This decision affects the degree of elitism or technocracy of the developmental effort.)
- Which social classes or interest groups will be made to bear the costs of change, and how will relative burdens be assigned?
- Which time spans are to be deemed tolerable before targeted gains are effectively reached?
- What degree of coercion from above will be judged acceptable?
- What measure of self-reliance or dependence on the outside is permitted or encouraged?
- Is priority given to material or to moral incentives? Or if to a mixture of both, in what proportions?
- Will the organizing principle of mobilized social effort be some form of socialism, a variant of neocapitalism, or novel indigenous approaches distinct from both?

The answers to these questions form a systemic whole which constitutes, in effect, a society's development strategy. Although decisions taken on all points are important, it is pedagogically useful to focus on a few of them, specifically: integration with outside systems (the international market or big powers), the degree of autonomy and self-reliance preferred (both as regards outside societies and, internally, within regions and classes of population), choices of overarching incentive systems, and the role played by austerity in formulating policy. These are now examined in detail.

Integration with Outside Systems

No Third World nation can successfully pursue a fully autarchic course. Even China, in spite of its continental size and a high degree of economic self-sufficiency, could not exclude all contact with the outside world.[6] Similarly, although Burma in the 1960s adopted a policy of excluding foreign investors and tourists, the country still needed to export rice and oil to outside markets.[7] Besides, nations like Tanzania and Sri Lanka have not interpreted the doctrine of self-reliance to mean exclusion of ties to other countries, to international agencies, or even to world markets. Nevertheless, important differences of degree are discernible among nations; some are more highly integrated with outside systems than others. At times, links are one-sided, as in the case of Cuba and its ties with the Soviet Union and Eastern European socialist nations. Algeria, on the contrary, provides an example of wide diversification, deliberately sought, in its linkages with other countries and regions. In Brazil, the primary integration sought is with the international market, not with a single nation or region. Preferences as to kind, degree, and locus of integration constitute a strategic development option fraught with consequences for technology policy. Indeed one major thesis propounded by Latin American *dependencia* theorists[8] states that domestic constraints on successful development are mainly due to structures of dependency imposed by outside forces operating in symbiosis with interest groups representing the social forces of "internal colonialism."[9] Whether one endorses or rejects these views, they correctly note how decisive for development strategy are decisions about degrees of integration with outside systems. It is no accident, therefore, that the "Brazilian model" of development imposed by the military government since 1964 has profoundly affected the kinds of technologies adopted, the allotment of social costs in the country, and the relative neglect of all but large-scale agriculture subsectors.[10] Once Brazil's planners decided to compete in the world market on the market's own terms, they were automatically giving direction to their strategy on other fronts and foreclosing alternative options.

By tying its development fortunes to an industrialized power or by striving to achieve competitiveness in world markets, a nation commits itself, often irreversibly, to certain industrial priorities or to large-scale industrialized agriculture over and against other alternatives, to certain patterns of consumer-goods production favoring the privileged classes, to supplying export needs over meeting internal demands, and to other policies having greater or lesser impact on technological choices. This is so because the "rules of the game" set by the world market or by the hegemonic big powers are biased in favor of obtaining development benefits in modes of large-scale competition, orientation toward higher purchasing power, and of a rapidly shifting "competitive edge" on the strength of changing technology.

More importantly, the decision to seek such integration necessarily relegates the concern for social equality to second rank. That the international economic order was designed to favor the already prosperous is conceded by no less moderate an observer than Gunnar Myrdal, who writes that

> the theory of international trade was not worked out to explain the reality of underdevelopment and the need for development. One might say, rather, that this imposing structure of abstract reasoning implicitly had almost the opposite purpose, that of *explaining away the international equality problem.*[11]

The existing global economic order is uncongenial to the pursuit of equity and equality because its wheels are lubricated by forms of competition founded on comparative efficiency. And using the capitalist, neocapitalist, and even socialist calculus of efficiency (to the extent that the latter "competes" in the world arena), such values as equity and equality are necessarily treated as "externalities" not to be "internalized." Therefore, whenever a national development plan in some poor country requires a high degree of integration with the global or regional export market, a whole gamut of supportive infrastructure investments is *ipso facto* rendered necessary so as to assure competitive efficiency.[12] Choosing integration implies selecting technology which is capital-intensive and of standardized international quality. It also signifies plant scales opposed to the requirements of small and medium industry as well as an agricultural policy which favors small minorities within the agricultural sector to the detriment of the poorest and least productive. Implied also are an employment policy which provides training and subsidies to small numbers of skilled and professional personnel—to the neglect (at least relative) of large numbers with lesser skills—and monetary and fiscal policies ill-suited to produce equitable redistribution inasmuch as subsidies favor "efficient" export sectors. For these reasons developmental efforts aimed at integration to big powers or to the world market set limits to strategies which can effectively be adopted. The recognition by Third World leaders that the present international economic order (IEO) is biased in favor of such integration helps explain their insistent demands for a new order.[13] Their interest in the international order is dictated by a recognition of the enormous impact their links to the world system have on their own domestic economies and policies. That is to say that their technology policies are directly affected by basic options regarding the degree and nature of these ties.

Integration with outside systems often means reliance on foreign "aid" as well as a commitment to produce for world-export markets. And linkages through "aid" are as crucial to technology policy as linkages through trade.[14]

In his speech to the United Nations Conference on Trade and Development in 1972, Robert McNamara urged the adoption of development strategies which make a frontal attack on the 40% of the poor world's population (800 million people, out of a Third World population of two billion) who have not benefited from past developmental growth or progress.[15] World Bank loans, he argued, should be granted if they reach this strategic portion of the population in poor countries. Dudley Seers later raised the spectre of rising unemployment as a further dimension of "failed" development.[16] Since then many strategists speak as though reducing poverty and creating jobs should be the core of their development strategies. Among those who most insistently plead for such strategies is Mahbub ul Haq, cited earlier. Recognizing that even expanding "modern sectors" cannot absorb a traditional sector whose absolute numbers grow ever more rapidly, Haq concludes that the inequalities generated by modernization strain "the limits of tolerance of many societies." He then asks why, if a dual economy exists, a dual development strategy should not likewise be formulated. His recommended strategy operates in two arenas:

> On the one hand, a modern sector which grows fast and experiments with all kinds of price incentives and tolerates the prevalence of inequalities for some time. On the other, a large traditional sector where organization and institutional framework overcome the scarcity of capital and development is taken to marginal men through the organization of rural and urban works programs.[17]

Haq wants greater self-reliance by poor nations in choosing development paradigms. His emphasis is valid because the nations which adopt an autonomous model of development, founded on optimal degrees of self-reliance, are also the ones most likely to insist on alternative approaches to technology.

Autonomy and Self-Reliance as Strategies

Tanzania holds pride of place among nations advocating self-reliance as a development strategy. One constant principle in the thought of President Julius Nyerere is that "there is no model for us to copy." He writes that

> in 1965 Tanzania adopted its own form of democracy—we rejected the Western model and said it was not appropriate for our circumstances despite the fact that all our constitutional development had until then been based on it. . . .
>
> When we introduced this new system, we were criticized for 'abandoning democracy.' . . . In response to this criticism we tried to explain what we were trying to do and why we thought our new system was both democratic and suitable for our conditions. But having done that we did not worry about what the Western countries said or what democratic theorists said. For in rejecting the idea that we had to follow the 'Westminster model' if we wanted to be democratic, we had also overcome the psychological need to have a certificate of approval from the West in relation to our political system.[18]

What Nyerere claims for a political system—freedom from servility to previously existing models—he likewise urges upon his nation in its approach to economic problems:

> We have deliberately decided to grow, as a society, out of our own roots, but in a particular direction and towards a particular kind of objective. We are doing this by emphasizing certain characteristics of our traditional organization, and extending them so that they can embrace the possibilities of modern technology and enable us to meet the challenge of life in the twentieth century world.[19]

The economy must be organized so as to free people from manipulation by the market; "the first priority of production must be the manufacture and distribution of such goods as will allow every member of society to have sufficient food, clothing and shelter, to sustain a decent life."[20] A bond is forged between self-reliance—in defining goals and in setting priorities—and development strategy. At a state banquet honoring Chou En-lai on 4 June 1964 Nyerere declared that both China and Tanzania are engaged in a revolutionary battle against poverty and economic backwardness. He added that for Tanzania the "long march" is economic. Other nations may learn from China that success requires not only courage, enthusiasm, and endurance but also discipline and as well the intelligent adaptation of policies to the needs and circumstances of each country at a given time. This is the heart of "self-reliance": the commitment to creative innovation and adaptation in the light of local constraints, values, priorities, and heritage. Any nation pursuing a self-reliant strategy of development must institutionalize its critique of prevailing outside models, capitalist and socialist alike. It must also adopt criteria for choosing technologies and modes of their utilization drawn from outside the technological market place. Foreign technologies are not excluded on principle, but imported will be only those types of technology which foster locally defined goals. Although it is violated by modern technology, nature is still deeply respected in rural localities. Schumacher contrasts the "self-balancing, self-adjusting, self-cleansing" attributes of nature with technology which recognizes no self-limiting principle as to size, speed, or violence. He concludes that "in the subtle system of nature, technology, and in particular the super-technology of the modern world, acts like a foreign body, and there are now numerous signs of rejection."[21] Critics view industrial societies as no model for export;[22] accordingly, they experiment with new modes of small-scale "community technologies" which enhance self-reliance in small local groups.[23] They teach that a basic option for self-reliance impinges on one's stance (if one is consistent!) toward technology. This lesson, in my view, is as valid at the level of national policy-making as it is at lower levels of decision-making.

The quest for greater self-reliance at the national level is the exclusive prerogative neither of small poor nations nor of those which have chosen socialism. Even giants such as the United States trumpet their wishes to be self-reliant in meeting their energy needs within a few years. Their desires are prompted by the need to reduce the kind of vulnerability any country experiences as a result of its integration with outside market forces. To be more precise, the aspiration after self-reliance in development models, in sources of capital or of technology, is motivated by the desire to reduce unacceptable forms of dependency. If the prevailing mode of exchange with others is interdependence with high reciprocity, there is correspondingly less "need" to be self-reliant. But if interdependence is characterized by differential strength or bargaining position (that is, by low mutuality or reciprocity) then a higher degree of self-reliance becomes desirable.

One important difference, however, distinguishes the self-reliance sought by groups within the US from the *basic option* to pursue self-reliance as a major mode of obtaining development (the path taken by China and Tanzania). In the latter's special circumstances, emphasis is placed on small-scale technologies geared to rural activities,[24] whereas in China a broad spectrum of criteria for acquiring technology is operative, leading to a "mixed" policy of combining large-scale "high" technologies in capital-goods industries with small-scale, locally improvised technologies in consumer-goods industries. China serves as an exceptional example of a nation adopting a developmental strategy which emphasizes self-reliance not only at the national level but also within regional, local, and productive-unit levels. Nevertheless, even self-reliance cannot be an absolute principle, and it must not be interpreted to mean excluding outside influences. Moreover, even where national planners do not choose self-reliance as their primary policy, it is possible within limited sectors (industry, let us say, agriculture, or housing) to champion a self-reliant approach which will have important repercussions on choices and modes of technology. This seems to be the context within which India has encouraged local initiative to create small technologies in industry.[25] Only in recent years have national-development planners begun to integrate science and technology policies with their central value choices and the corresponding development strategies. In the past, technology policies were either abandoned to the workings of existing technology market channels or dictated by local or sectoral actors in economic decisions. Thus if a ministry of housing favored unaided self-help public housing utilizing cheap technologies, sectoral policy was shaped with little regard for macroeconomic options. The degree of centrality which integration to outside markets or self-reliance as an organizing principle of development effort assumes is crucial. Where integration is the basic option, higher degrees of conformity to technological patterns dominant in industrialized countries are unavoidable. Where, in contrast, self-reliance is the primary mode of development, greater leverage exists for reducing such conformity. Self-sufficiency doubtless has to be be paid for by sacrifices in efficiency; conversely, higher integration, although it may improve efficiency, may exact sacrifices in terms of social justice and lead to excessive industrial concentration or vulnerability to price fluctuations over which national decision-makers have little control.

Other things being equal, if a less-developed country subscribes to the image of a procession of nations struggling to "catch up" to the developed, it will *ipso facto* be influenced by a powerful bias in favor of choosing integration (via aid, trade, imported technology, and the adoption of international standards) as its basic option. If, on the contrary, the primal image it adopts is one of revolutionary convulsion, the antecedent probability exists that self-reliance will hold an important place in its basic development strategy. Practical constraints of a military, economic, or political nature may sometimes overrule this preference, as in the case of Cuba, which chose to become highly integrated—militarily, financially, politically, and technologically—with the Soviet Union. The third image, multiple apocalypse, tends strongly toward the basic option of optimizing, at all levels, both self-reliance and control over growth. Similarly, if one takes as a basic strategy a frontal attack on mass poverty and unemployment, a strong bias exists in favor of lesser integration with international markets and greater local inventiveness to correct factor distortions inherent in technology imported from rich countries. In every case there are limits beyond which neither efficiency nor equity can be fully ignored. Yet on balance, a basic option on the scale of degrees of integration and self-reliance is the central parameter within which technology policies can be evaluated. Because reality constantly imposes compromises, no country plan is fully consistent with its basic options, and unexpected events (such as abrupt rises in import prices or disastrous floods) can suddenly make relatively self-reliant nations more "integrated" with the outside than formerly. Nonetheless, any nation's decision-makers will ultimately have to attach primary importance either to integration or to self-reliance. The degree to which they blend the two will depend, in large measure, on the overarching incentive systems at work in their societies.

Overall Incentive Systems

The category of "pure types" devised by Max Weber for heuristic purposes helps explain what is meant by incentive systems. The market is a vigorous mechanism for motivating people to produce, sell, buy, and consume. It presupposes the existence of a shared incentive system around which buyers and sellers visualize comparative and mutual advantages in playing out their respective roles. The existence of effective purchasing power in the hands of a pool of prospective buyers is the incentive leading producers and consumers alike to engage in the dynamic processes of generating supply and demand. In market "societies" (a more accurate term than market "economies"), therefore, the *organizing principle of economic effort* is the market. The market undoubtedly needs various subordinate mechanisms to work—price, competition, and demand stimulation. Nevertheless, the market itself constitutes the overarching incentive system. Even a conservative economist like Henry Wallich, as he reflects on worldwide trends towards more socialism and state ownership of productive enterprises, unhesitatingly affirms that the survival of capitalism and the continued profitability of large corporations depend not on continued corporate or private ownership of productive assets but mainly on continued access to markets. Wallich is convinced that "a market-oriented system can operate with any kind of ownership—private, public or mixed."[26] Most criticisms leveled at capitalism, he adds, should be addressed not to the market, which is merely one (albeit a vital) institution within that system, but rather at private ownership, which, he alleges, is the cause of growing inequalities.

Indirect confirmation of the functional role of the market as overall incentive system comes from the initiatives taken in recent years within the United States to institute "political marketing."[27] The key to the concept of political marketing is the belief that the purchasing power of socially conscious people can and should be harnessed to help the poor economically, not just politically. Effective purchasing power unmistakably serves as the basic incentive system of economic life in capitalist societies. It must not be supposed, however, that societies organized around other basic incentive systems can dispense with markets. On the contrary, even socialist societies rely on markets. But instead of serving as the organizing principle of economic life, the market in these societies acts as a regulatory mechanism to control against waste, duplication, and overcentralization. The market is subordinated to another organizing principle: a plan relying on varying degrees of coercion or persuasion, some consensus about priority needs, or pervasive mobilization around collective moral incentives. Writing in 1951, Karl Mannheim explained the difference between an organizing principle and a subordinate mechanism:

> Competition or co-operation as mechanisms may exist and serve diverse ends in any society, pre-literate, capitalist, and non-capitalist. But in speaking of the capitalist phase of rugged individualism and competition, we think of an all-pervasive structural principle of social organization. This distinction may help to clarify the question whether capitalist competition—allegedly basic to our social structure—need be maintained as a presumably indispensable motivating force. Now, one may well eliminate competition as the *organizing principle* of the social structure and replace it by planning without eliminating competition as a *social mechanism* to serve desirable ends.[28]

What Mannheim says of competition is true as well of the market, the matrix which legitimates economic competition. Within socialist countries the market is subordinated to planning, which serves as the economy's organizing principle or overarching incentive system. Competition, therefore, is not eliminated but takes on secondary importance: It is not the dynamic motor of effort; it merely stimulates and channels effort, usually under the rubric of "socialist emulation" and in support of targets set by some plan. For example, Fidel Castro repeatedly urges Cubans to maintain some balance between moral incentives (expressed as solidarity with the neediest and the "mission" to build socialism) and material incentives (expressed as desires to improve one's material lot). Citing Marx's view that "rights can never be more advanced than the economic structure and the cultural development determined by it," Castro warns against basing labor and sacrifices too exclusively on either material, or on moral, incentives:

> It is true that many of our workers are real examples of Communists because of their attitude toward life, their advanced awareness and their extraordinary solidarity. They are the vanguard of what all society will one day be like. But if we think and act as if that was the conduct of every member of society, we would be guilty of idealism and the results would be that the greatest share of the social load would unjustly fall on the best, without any moral results in the awareness of the most backward, and it would have equally negative effects on the economy. Together with moral incentive, we must also use material incentive, without abusing either one, because the former would lead us to idealism, while the latter would lead to individual selfishness. We must act in such a way that economic incentives will not become the exclusive motivation of man, nor moral incentives serve to have some live off the work of the rest.[29]

Castro broadly equates idealism with moral incentives and individual selfishness with material incentives, but of course each category admits many variations. Idealism may take shape in an effort at nation-building, or the creation of a socialist society, or in a mobilization directed toward revitalizing ancient cultural values threatened by "materialistic modernity." Material incentives likewise admit multiple emphases: they may be *personal*, founded on what McClelland calls "achievement motivation,"[30] or *collective*, expressed as a desire to "catch up" with the West in steel production or space programs or the range of consumer products manufactured locally. The Cuban leader correctly concludes, however, that no incentive system can dispense with a mix of both material and moral elements. The crucial questions are the primacy given to one over the other and the specific values which are appealed to as mobilizers of social effort.

Under any development strategy, great attention must be given to educating masses, specialists, and leaders in the duties of *solidarity*. Yet one must still ask: Solidarity for what? At times solidarity aims at harnessing the energies of all in order to abolish both absolute poverty and scandalous inequalities among persons and social classes. Recent studies show that few countries have adopted this as their priority or have achieved great success.[31] Again, one notable exception is the People's Republic of China, which deliberately set out to meet basic needs of all the populace in a highly egalitarian manner.[32] Not only does that state guarantee employment to everyone who seeks it, it also provides health, education, and transportation to all at nominal costs. Income policy deliberately minimizes differences in wages. Most visitors, including even those who are unsympathetic to China's ideology, have reported that no one is hungry, unemployed, or bereft of basic health services in that land.[33] The elimination of mass famine, endemic disease, and unemployment in China would have been impossible in the absence of an overarching incentive system anchored in moral solidarity and requiring all to contribute for the benefit of all.[34] The ideal is to achieve what one author calls "a modest but fair livelihood."[35] Personal entrepreneurship is condemned as bourgeois selfishness and, though much stress is placed on working productively and rapidly increasing the available stock of goods and services, equal insistence is placed on producing more for the benefit of all, especially those in direct need. Nowhere is social struggle so frequently and so prominently invoked as pedagogically necessary as in China; the notion of "walking on two legs" is a recognition of the need to live in tension between conflicting demands. The choice of a moral incentive system to lend dynamism to its development effort constitutes, therefore, a basic option taken by China which effectively commands development strategies in domains of investment priorities, locational decisions, and modes of enterprise administration.[36] Successful policy in domains such as health, population, education, and transportation, on the other hand, is conditioned by the incentive systems which preside over mobilization efforts made by the nation struggling to achieve development. Certain financial, material, and labor input become possible only if moral incentive systems prevail. Conversely and in other societies, certain policies can be adopted only where the overarching incentive system allows market competition to favor individual or corporate enterpreneurs and consumers.

As noted earlier, no incentive system can totally dispense with some mixture of material and moral goals. Even within market capitalism the assumption is usually made that inequities will be offset either by mechanisms of corrective state distribution, by formal and informal policies constituting some form of charity, or, more frequently, by the putative "trickling down" of benefits from the rich to the neediest. This is generally the case even where policy-makers accept the view of Kuznets that initial periods of development produce wider income disparities than previously existed.[37] One difficulty with any trickle-down vision is that incentive systems, like other societal mechanisms, are self-reinforcing: they develop inertia and a momentum of their own, usually reinforcing initial trends. Myrdal long ago acquainted students of development with the "vicious circle" of poverty, an image which reappears in his later writings. There exists, analogously, a vicious circle of reinforcement of vested interests at work in overall incentive systems. Once an economy is organized in response to the stimuli of individual achievement, it becomes extremely difficult to introduce a new catalyst into the system without having it neutralized. An illustration is found in the ease with which US corporate officials interpret their "social responsibility" (in domains of ecological integrity and social justice) in terms of profitability. The president of Ford Motor Company's Asia-Pacific division writes that

> Ford Motor Company believes in the profit motive. But we do not see corporate profitability and corporate social responsibility as mutually exclusive. As our Chairman, Henry Ford II, said recently: "A corporation can serve society only if it is profitable."[38]

A more striking expression of this view comes from Carl A. Gerstacker, chairman of the Dow Chemical Company: "It is in reality the profit motive that makes industry responsive to social needs."[39] Therefore, once it becomes "profitable" to be just, managers moved by "material" incentives will respond. What is left unsaid in such declarations is that the victims of injustice lack "effective purchasing power," which is the primary stimulus to which profit-seekers respond. Clearly governments may act as intermediaries between profit-makers and needy consumers bereft of buying power. Yet even equitable social-welfare policies must overcome the dominant influence exercised on sectoral policy by the basic incentive systems. Wealthier taxpayers, who provide the funds governments use to play their welfare role, must at the very least perceive the gains to themselves as worth the cost. Not surprisingly, powerless needy groups are neglected while potentially dangerous (read: embryonically powerful) ones are appeased. The perdurability, within industrial countries, not only of "pockets of poverty" but also of numerous social groups "left out" of the putative benefits of development suggests that neither trickle-down nor governmental correctives to the "invisible hand" of the profit incentive can abolish mass misery. Champions of capitalism usually claim that moral incentives lead to excessive sacrifices in efficiency.[40] There is doubtless some truth to the claim. Yet no development strategy can avoid making a basic choice: either to "develop" for the benefit of those who are already privileged or who constitute promising potential candidates for "modernity" or to embrace a pattern of "development" aimed at abolishing dehumanizing misery for all its citizens and to create an incentive system capable of institutionalizing this priority. Some form of "walking on two legs" in inevitable. The almost universal failure of trickle-down and welfare statism suggest, at least tentatively, the greater promise lying in granting primacy to moral incentives. All strategies, however, will fail without suitable doses of imposed austerity.

Austerity: The Price of Technological Freedom

Few words and policies are as unpopular as *austerity*, a term which evokes the notion of involuntary sacrifices, usually imposed on those whose initial level of welfare is lowest. Almost always austerity policies are announced as temporary evils which will make future prosperity possible. But, as P.T. Bauer observes,

> current austerity does not in the least ensure future abundance and does not even generally promote it. Indeed, policies of current austerity tend to perpetuate it, in a number of different ways: by reducing the supply of incentive goods; by divorcing output from consumer demand; by politicizing economic life; by provoking political tension; and in various other ways as well.[41]

Nothing is gained by claiming, for an austerity policy, more than it can deliver or by assuming that it operates benevolently. But austerity works only when it does precisely what Bauer condemns, namely, when it "politicizes economic life." The vital issue is *how*. Chinese policy-makers characterize austerity as an attitude of "bearing up and striving on."[42] For the rulers of contemporary Chile, on the contrary, austerity conjures visions of desperate measures applied only to forestall total economic chaos. Unlike its Chinese version, designed to assure high degrees of equality among the entire populace, the Chilean concept of austerity is designed to keep inflation down and to spur economic investment, even if this means imposing heavy burdens on poorer classes and favoring foreign investors.[43] Austerity can be imposed for diverse reasons and to the advantage, and disadvantage, of quite diverse social groups. An important difference is to be found in the underlying imagery: austerity seen as a necessary evil or as a permanent component of developmental humanism. The former conception is purely instrumental: under certain conditions austerity or belt-tightening is accepted as undesirable but necessary. But the second view considers austerity as a value for its own sake, even if it is not strictly necessary on purely economic grounds.

The conventional view holds that during the early stages of capital accumulation—when initial levels of material well-being are low—austerity, or the limitation of consumption, even of quite basic goods, by the masses is a "necessary evil," an unpleasant but unavoidable sacrifice required if capital is to be saved in order to be invested in productive facilities which will generate future increases in consumer goods. This conception says nothing about criteria for assigning the social costs of austerity to specific social classes or interest groups. But in fact it is often associated with a policy of favoring large investments and providing incentives to wealthy investors; predictably, therefore, it exploits the poorer elements in the populace. China's notion of austerity, is diametrically opposite: Austerity, or the willingness to be content with a decent sufficiency of goods, is viewed as a permanent component of authentic socialist humanism, because a certain detachment from abundance, whether presently enjoyed or desired in the future, is considered the necessary basis for establishing the primacy of moral, over purely material, incentives.[44] Without such moral incentives, rooted in solidarity and struggle to improve the lot of "all the people," the acquisitive spirit will impede the revolutionary task of "building a new man and woman." Furthermore, the Chinese believe that one's desires are alienated in the vision of future affluence no less than in servile clinging to present possessions; thus, both postures are inimical to struggling on behalf of egalitarian gains. Mao, fully conscious of the "heretical" nature of his teaching, condemned the Soviets for climbing aboard the "goulash and television bandwagon." The Chinese concept of austerity, however, is the very antithesis of "tightening the belts" of the poorest. On the contrary, it preaches sufficiency for all and places the vital needs of the masses in higher priority than luxuries for the few.

This notion does not make a fetish out of deprivation; it is understood that as conditions improve, material levels of comfort, utility, and enjoyment should rise proportionately. But they should rise in a manner consonant with two values more important than growth: social equality and the struggle against technological determinism. The link between austerity—whether imposed or voluntary—and "technological freedom" is central.[45] A deliberate value decision not to pursue affluence in goods and services—or at least to subordinate that pursuit to other considerations—is an indispensable means for obtaining control over technology. Jacques Ellul judges that "as long as man worships Technique, there is no chance at all that he will succeed in mastering it."[46] No society and its members can avoid worshipping technique unless they practice austerity, not as an instrumental "necessary evil" but as a permanent component of authentic developmental humanism. As stated earlier, technology creates its own momentum around the alleged need to keep making new products and devising new processes. This momentum fuels competition: among firms, to gain markets; among nations, to achieve military or geopolitical supremacy, thanks to technology. Or the competition may take a different form and be waged by poorer countries against comparative indices or "more modern" societies. Powerful determinism." This is the argument of classical Marxists, according to the other hand, as some have argued, the deliberate quest of "technological freedom" can be internalized in one's cost-benefit choices at the outset so as to minimize this "tendency towards technological determinism." This is the argument of classical Marxists according to whom technological alienation is traceable to capitalism and exploitative relations of production, not to any trait inherent in technology itself.[47] Others, operating on different philosophical premises, insist that "people's control" over technology becomes a major objective sought in the initial choices made of technology.[48] They favor small-scale, "soft" technologies. The point is, simply, that *resistance to determinism* must itself become an explicit feature of all technology policy and that a direct correlation can be found between one's viewpoint on austerity and the possibility of countering technological determinism. The possibility of such a link can perhaps be adumbrated by suggesting several levels at which prior choices regarding the pursuit of affluence or the nurture of healthy austerity imply divergent technological policies. The problem may be discussed first in terms of national policy.

National-development planners who initially opt in favor of achieving the whole gamut of industrial capacity, both for a domestic and for foreign markets, cannot avoid a high degree of dependency on outside suppliers of technology. Some countries, such as Algeria and Brazil, view this dependency as only temporary; they express confidence in their ability to import technology on a large scale and from diversified sources for a limited number of years and, in a second phase, to gain a relative degree of autonomy. Only the passage of time, perhaps two or three decades, will reveal whether their sanguine expectations are realistic. What is certain, however, is that even if LDCs achieve competitive position, thanks to a policy of rapid technological purchases, allied to an intensive program of training nationals abroad and of inaugurating technical institutes at home to create indigenous R&D capacity, they will not necessarily be free from technological determinism. Under favorable circumstances, perhaps a few Third World nations can acquire relative technological autonomy from outside sources, but in order to do so these nations must enter the competitive arena of technology. And once they have entered they will be obliged to change constantly in order to maintain their own "competitive edge." By integrating their economies with the vicissitudes of world markets, they will fall prey to irresistible pressures to achieve constant technological change. In short, *gaining technological competitiveness is not synonymous with gaining technological freedom.* The freedom in question is that of exploring technologies better adapted to local needs and of creating conditions which allow a society to control, at least relatively, the speed and direction of technical change. To these realms must be added the freedom *not* to become enslaved to mass-consumption models of development, which perpetually call for new products, new packaging, new processes, and new markets. The relationship between technical freedom and austerity is, in a word, *direct.*

Societal control of technology is a value to be *internalized* in technology policy, not treated as a mere *externality.* And some form of austerity is needed in the very determination of efficient production inasmuch as high priority must be given to meeting the basic needs of the masses over satisfying the wants of those already enjoying high purchasing power. A philosophy of austerity embracing the goal of "sufficiency for all" will favor technologies which do not foster widening gaps in income, in degrees of sharing in decisions, and in class stratification. The achievement of these objectives lies at the heart of technological freedom. Accordingly, a deliberate option in favor of austerity, with its attendant emphasis upon self-reliance, might lead nations to do without certain technologies if these can be acquired only at intolerable costs in money, in dependency, or in conflict with equity goals. Abstention from imports can encourage local and regional innovation, using less costly materials, less highly trained personnel, and more readily mastered techniques. To the objection that this approach condemns a nation to a subordinate role on the world scene, it must be answered that the "high technology" strategy does not abolish misery among masses, create employment, or facilitate genuine development. Of course, even a sound austerity policy cannot be applied absolutely or unidimensionally. Nor need it exclude the selective importation of modern technology, provided this activity is subordinated to larger goals. So long as development was equated with aggregate economic growth, with little regard for distribution of benefits or for the creation of jobs, or *a fortiori* for the creation of decisional democracy in other than purely token ways, technological dependency could be viewed as the "necessary price" to be paid for genuine benefits. However, now that development thinkers themselves are questioning the merits of the purely growth-oriented version of development, technological dependency appears in a new light. If, to the disadvantages just named, one adds those relating to ecological damage, the argument against unrestrained technological imports gains added weight. To consent to do without certain foreign technologies is tantamount to doing without certain consumer goods, services, and capital goods usually associated with "abundance" or "affluence." Implicitly, it means an internal production strategy of "austerity." The central decision bears on the nature of goods produced, inasmuch as production determined which technologies are employed. Viewed in this light, "austerity" is no longer the business of elite decision-makers "tightening the belts" on the already poor but the economic expression of a society's commitment to placing the needs of all above the wants of the few. If such an option is to be freely accepted by a populace, a certain pedagogy of austerity will clearly be required. Yet no "Madison Avenue campaign in reverse" can succeed, inasmuch as advertising itself is a leading manipulative tool.[49] Pedagogy must be conducted by *persuasion* as to the importance of not falling prey to the blandishments of technology and wasteful consumerism. If such a pedagogy is to prove successful, it must obviously not be conducted at the purely individual level. Like other forms of political marketing, it has unmistakable social dimensions. Consequently, government policies at the macroeconomic level must be designed to support frugality and the war against waste.

Austerity understood as "sufficiency for all as first priority" is the only path which can directly attack the poverty of the poorest majorities. All other measures can have but palliative effects at best. This view formally repudiates all versions of the trickle-down theory, according to which material improvement of the poorest can come only from a growth in wealth. It is also opposed to conventional distributive theories which concentrate mainly on reapportioning wealth *downstream*, that is, without affecting basic ground rules governing access to resources *upstream* (before they have been exploited or processed). The disappointing results of endless discussions regarding a new regime of the seas and of seabed resources suggest the near impossibility of applying technology to creating new systems of resource equity without a prior commitment to the principle of sufficiency for all as first priority.[50] One major problem is that the wielders of technology are frequently the beneficiaries of the system that assigns first priority to those enjoying the greatest purchasing power. Industrial and managerial technology in their present forms are not designed to foster austerity. Therefore, a rupture with transfer systems must be made on grounds of austerity if a poor society is to escape technological thralldom.

And why should an austerity policy in technology not be applied absolutely? Because even countries with a dearth of resources will need advanced technology in certain domains; they need a judicious choice of many technologies. Not every country with a small market can be self-sufficient even in the production of modest consumer goods. But if these societies place a high premium on the twin freedom from technological dependency on outsiders and from technological determinism tied to competition in the ever-shifting race after new goods, austerity becomes unavoidable. The message of these pages is not that austerity is desirable but that it is *necessary if* technological freedom is valued. Austerity constitutes one of those basic options which, if pursued coherently, sets limits to development strategies and to specific sector policies. This is why we need to challenge the several conventional theses that austerity

- is necessarily harmful to the poorest sectors of a populace
- cannot provide incentive systems which stimulate rapid gains in production and productivity
- cannot be compatible with controlled progress in high technology itself
- needs to be imposed by coercion, rather than accepted by the population at large

If exercising human control over technology is a monumental problem facing humanity today, it makes good sense to seek material improvement and institutional and structural transformation with an eye to minimizing the momentum inherent in technology. The powerful stimulus of competition will not then be abandoned but subordinated to deeper values—equity, control, and balance. These will now command technology instead of serving it.

8/Third World Technology Policies

Technology is "a powerful means of international policy: it serves as a new means of projecting national influence and power into the international arena."[1] Consequently, a growing number of Third World scholars advocate technology policies which promote not only development but greater national autonomy as well.[2] At conferences on science and technology one hears laments over the absence of an explicit technology policy in many countries. The 1972 OAS meeting of Latin American governments in Brasília declared that

> science and technology offer infinite possibilities for providing the people with the well-being that they seek. But in Latin American countries the potentialities that this wealth of the modern world offers have by no means been realized to the degree and extent necessary.[3]

The "eminent persons" testifying before the United Nations on the role of transnational corporations in technology transfer argue that political control of such firms is the heart of the question.[4] Control is sought through special legislation, a single component in a broader national policy. This chapter explores policies consonant with the aspiration of many Third World nations to acquire greater technological maturity, viability, and autonomy. The domains covered by technology policy will be mapped out briefly. Then the constitutive elements of an appropriate technology policy are discussed: strengthening infrastructure, perfecting negotiation stances, launching two-way technology flows, taking concerted action, among Third World actors), and seeking appropriate forms of support from international agencies.

Arenas of Technological Policy

A careful formulation of technology policy is as important as development-planning itself. Good plans can be brought to naught if technology is insufficient in quantity, inappropriate in quality, or undisciplined in its applications. And given present threats to planetary survival, to ecological integrity, and to equitable resource use, one can no longer "cheat" on these issues.[5] Developing nations cannot afford *not* to have a technology policy. To have no policy is to have a bad one; technology affects development too crucially to be left to its own momentum. Increasingly this momentum is being viewed as destructive, a critical trend which prompts John Montgomery to conclude that "the Neo-Luddites are protesting, not against machines, but against the technology that produced them and keeps them running."[6] Technology is seen by many as the source of exploitation. A recent political manifesto declares that

> we are in the midst of a gigantic revolt against the effects of the North Atlantic technological civilization, against the inability of civil society to harness technology to human ends. This revolt is the fundamental cause of the conflict; it is at the root of most international and domestic problems.[7]

At a time when most Third World countries demand more technology, decision-makers need to clarify and bring some measure of coherence to their policy objectives. Most poor countries postulate five basic goals to be served by their technology policy:

- to obtain the entire gamut of available technologies
- to optimize the use of those technologies within their societies in ways supportive of basic value options, development strategies, and policy determinations in various sectors
- to create and disseminate as widely as possible a technological culture or mentality
- to build up the capacity to produce their own technology
- to have a fair pricing structure for their technological imports

These are general goals, but specific instrumentalities for reaching them focus on other issues. Technology policy deals directly with the supply, demand, and role of diverse actors in the circulation of technology. On the supply side, policy-makers seek to build up the nation's technological base by measures such as investment in science infrastructure (teaching of science, research, promotional incentives to recruit and retain personnel), reshaping the general educational capacities of their societies, and other measures aimed at creating a technological frame of mind in the populace at large. These are the realms where policy affects technology supply.

What complementary measures touch on the demand for technology? Most generally, the aim sought by concerned governments is to rationalize their demand system by organizing information on technology's availability, scope, and character;[8] providing incentives to stimulate demand for local technology; establishing linkages among policy-making, production, and research; and regulating technology to make it serve preferred objectives.

No piecemeal approach to technology policy can prove satisfactory. Legislation fostering the creation of national technology may prove futile unless parallel efforts are made to integrate technology to producers in ways congenial to development policy. Similarly, incentives held out to users of technology to buy from local suppliers may fail if the supply pool is not adequate or available. The need for comprehensive policy has given rise to the interesting theoretical scheme named after Argentine physicist Jorge Sabato.[9] As discussed earlier (Chapter 4, pages 81-83), the "Sabato triangle" is both a tool for diagnosing the deficiencies of national technology structures and an instrument of prescriptive policy. Its imagery is simple: government policy-makers form one apex of a triangle, producers a second, researchers a third. Just as the "triangle" is made up of three points and of lines joining all three, so too a "Sabato triangle of technological circulation" requires communication flows linking every pair of elements in the system. Hence no good supply-demand system exists unless government policy-makers take producers' needs into account as they frame their research priorities. Conversely, producers must be intent on complying with policy objectives. Two-way flows of information, goal-setting, and problem-solving must also exist between researchers and producers, as well as between both of these and policy-makers. In many Third World countries, research is unrelated to problem-solving faced by industry and other productive sectors. To make matters worse, producers themselves are largely indifferent to targets set by national policy-makers. And more often than not, planners formulate no technology policy at all. Yet, as one important United Nations document notes, "if science and technology are to make an effective contribution to development, there must be a direct relationship between science and technology and government policy."[10] Policy is not limited to promoting science but embraces all issues relating science and technology to development. Sabato urges that working "triangles" be set up in all major sectors of the domestic economy, as well as in the technology-import sector. Otherwise, much waste ensues and technological needs remain unsatisfied or must be met at exorbitant costs. The successful operation of a triangle, however, is conditioned on the existence of some research infrastructure and, more importantly, an articulated network of information.[11]

One major complaint voiced by officials from poor countries is their difficulty in obtaining even nonproprietary information regarding available technologies.[12] One major service provided by technology answering services, such as those operated by the United Nations Industrial Development Organization, the Society for International Development, or the Volunteers for International Technical Assistance (VITA), is a reduction in the expenditures required of poor countries to obtain such information. A greater problem than lack of information is low absorptive capacity. A society must already have some capacity to create technology in order to be able to absorb outside technology creatively. Thus policies aimed at improving channels of access to foreign technology necessarily call for shoring up assimilative capacities within importing nations. A similar imperative is at work within the sectorial domestic "triangles": research enterprises must learn something of producer needs if they are to absorb "demand" information from industry in ways conducive to effective "supply" responses. It is useless to speak of the need for "infrastructure," however, unless one states what specific goals that infrastructure must serve. Like a scaffolding designed to support an edifice while the latter is under construction, so too social infrastructures must support larger systems distinct from them. Accordingly, one must ask how science and technology infrastructures can be strengthened to serve development purposes.

Technology Infrastructure

As early as 1958 George Counts asserted that "the most profound questions regarding the conduct of education, questions involving values and purposes, will have to be found outside the school and beyond the imperatives of scientific knowledge. . . . Technology has raised anew and on a vaster scale than ever before the ancient question of the values by which men live."[13] Counts is right: it is not the imperatives of scientific knowledge which will dictate the type of school or general educational infrastructure. Of course it makes no sense for a country to seek modernity unless it equips itself to obtain it. And modernity cannot be attained without technology. Perhaps imported technology in the past enabled less-developed countries to raise their production levels.[14] Throughout the Third World, however, the sentiment now exists that high dependency on outside suppliers of technology is an indignity. Quite apart from its economic and social costs, dependency psychologically affronts the consciousness of Third World nations which do not wish to be mere consumers of civilization. Consequently, many LDCs now seek to mount their own science and technology infrastructure, a step rich in prestige value. What realistic prospects, however, do most Third World nations have for gaining scientific and technological capabilities? Do not such capabilities lie beyond their means?

No flat answer to this question can be valid, for even relatively poor countries can strengthen their infrastructure. The adoption of sound curricula and modes of instruction can help introduce the "scientific mentality" among the general population. Under favorable circumstances it can even spread the experimental spirit among cohorts of the young initiated to laboratory work. Even the poorest societies have a certain number of professionals, including some scientists. The questions to ask are: What tasks occupy them? Are these tasks unrelated to technological problem-solving in their countries? What rewards or sanctions stimulate producers to look for local answers to technological questions? What inducements do governments offer to their nationals to adjust research to productive needs?

Whatever be the scale or initial conditions of scientific work, it is evident that some margins exist in all societies for improving their scientific and technological infrastructures. Admittedly, improvement is costly and often takes many years to become tangible, but significant progress is possible nonetheless. Planners need to identify key sectors of the economy wherein high degrees of technological dependency are intolerable—copper-mining in Chile, cocoa-growing in Ghana, cattle-raising in Argentina. Next they can inventory research capacities. Even if these are modest, they can probably be coordinated better to solve practical problems in the key sectors. The objective is simply to reduce gaps between directions taken by research and needs of producers. Vietorisz tells of a research institute with "a brilliant record of publication by its engineers and researchers in respected international technological journals, yet local industrialists never went near it, and looked toward international consulting firms when needing technical business advice. This is an instance of a total lack of integration between two kinds of institutions each of which is indispensable for technological autonomy."[15]

Proper linkage between potential suppliers and utilizers of technology is aided by incentives which reassure producers that answers to their needs can come from local researchers. In short, creators of technology must be stimulated to do what is relevant to producers, and utilizers must be induced to have recourse to national suppliers of technology. Because most Third World producers look outside the country to satisfy their technological needs, direct intervention by governments in the form of coordination, direct coercion, and subsidies to researchers is required to reverse the pattern. In several countries legislation requires firms to use national technologies if these are available.[16] In addition, national technology registers require that all technology contracts with foreign suppliers be recorded and approved. These are weak instruments, however, in the absence of an incentive system which meets the needs of firm managers. Restrictive legislation cannot force industrialists to contract national suppliers of technology unless the latter are able to make the former competitive in their primary markets. But neither registry nor restrictive legislation is meant to eliminate technology transfers. A recent essay in *The Economist* states: "The basic issue is not the one at which the experts were beavering this week, to control technology transfer. It is to transfer technology."[17] Even in state-owned firms, enterprise managers must meet production targets. Therefore, they look for technology delivered on time, reliably, and in ways which satisfy their customers. So long as national suppliers of technology cannot meet these needs, buyers of technology will lack compelling motives to stop relying on foreign suppliers.

National policies for strengthening technological infrastructure must comprise three elements:

- incentive systems to harmonize interests of purchasers and sellers of technology with national policy objectives
- informational inputs and legislative constraints to wean producers away from outside suppliers
- proper subsidies to enable national suppliers to meet the technology needs of national purchasers

The incentive systems discussed here are less general than those analyzed in the previous chapter, which focused on the organizing principles of societal mobilization. At issue in the present context is a coherent set of measures to meet what Vietorisz calls "the test of technological integration."[18] Technology is no mere commodity, but an overall system, and the test of integration is the extent to which a country is capable of technological autonomy. Even partial autonomy is out of reach until a pool of national technologists exists and is properly organized to meet clients' needs. To create or build up such a pool is the first priority in any plan to build up infrastructure. Information systems are also needed, to serve two functions: to join potential buyers and sellers of technology and to convince the society at large that native capacity can solve technological problems. According to Nigerian economist S.A. Aluko, "One of the main causes of technological backwardness is the lack of confidence among African leaders and governments in their own ability and that of their people to solve many of the local problems."[19]

Legislation is sometimes needed to entice national producers away from international technology suppliers to national ones. But there are limits to what can be accomplished by legislative infrastructures unsupported by operations to match up users with suppliers of technology. Thus, notwithstanding stringent legislation passed in late 1974 in Argentina prohibiting recourse to foreign suppliers of technical expertise when parallel skills are available within the country, in practice firms remain dependent on outside suppliers.[20] One reason is international prestige. The status which accrues to successful firms in the First World has spread to much of the Third World. The presumption is that what comes from the "developed" world is better than what is produced locally. Paradoxically, however, US consultant and engineering firms often send their less-qualified personnel to Third World sites, especially when these firms "spread themselves too thin" so as not to lose contracts. Hence Third World countries need to undo the myth of the technical superiority of "developed" outsiders. The best way to do this is to create professionally competent local counterparts. The success of local consultants in many domains attests to the feasibility of this strategy. National producers require technological help at various stages:

- pre-investment and feasibility studies
- engineering designs, machinery specifications, plant designs, factory layout

- equipment selection, plant construction, installation, and start-up of plant
- acquisition of process or manufacturing technology
- technical assistance during the postinstallation period, including training programs and management assistance[21]

Not all technological expertise can be developed rapidly, and there is a definite sequence of accessibility. The capacity to conduct feasibility studies may well be the last skill a Third World consulting group can acquire. Nevertheless, as Albertal and Duncan note,

> there are experienced consulting groups in Brazil and India; competent economic planners in Kuwait and Pakistan; able public managers in Ghana; well-trained fisheries specialists in Cuba and the Philippines; up-to-date industrial technology experts in Argentina; qualified forestry specialists in Malaysia; outstanding centers for administration in Costa Rica, Sudan and Venezuela; and in many countries, research institutions that could make unique contributions to developing countries. We also know that they are often under-utilized both at home and abroad.[22]

Greater use of Third World skills would boost local infrastructure considerably.

Great progress remains to be achieved in management technology which, according to some, "is the essential technology—the one upon which effective utilization of all other technologies must rest."[23] Management technology utilizes people so that they can in turn utilize all other technologies in optimal fashion. And management know-how is, by definition, attentive to local contexts. Consequently, local managerial capacity must be reinforced, if other technologies are not to be ill-used. This view is confirmed by the studies of John W. Thomas, who concludes that the most crucial variable in determining what kind of technology is adopted is the organizational capacity of the vehicle.[24] Thomas has collected evidence from Bangladesh, Tunisia, Turkey, and Ethiopia to support his contention that "in most situations the organizational structure of the agency undertaking the activity will dictate to an important degree the type of technology employed."[25]

Sound managerial infrastructure helps provide a necessary incentive system for harmonizing the interests of purchasers and sellers of technology with national policy objectives. It also sets up procedures assuring informational flows among the three "points" of the Sabato triangle and provides an institutional base for allowing needed subsidies to flow from policy-makers to researchers and producers. An acceptable infrastructure thus includes a pool of scientists and technicians, organizational and managerial institutions to assure linkage in the "triangle," proper legislation to counter the advantageous position presently held by many outside suppliers, subsidies and other incentives for national buyers and sellers of technology to deal with each other and with government policy-makers, and informational networks without which full optimization of local technologies and optimum use of imported technologies will prove impossible. These instruments are a prerequisite for a policy of appropriate technologies.

Appropriate Technologies

Great confusion surrounds the terms "appropriate," "intermediate," "soft," and "humane" technologies.[26] Discussion often implies that only labor-intensive, resource-conserving, and small-scale technologies are "appropriate" for poor rural nations. A philosophical preference for "gentle" technologies, however, is no helpful policy guide for selecting the technologies a nation needs. In short, *no single technology is "appropriate" for all developmental purposes, but every technology is "appropriate" for reaching some objectives.* Thus, the first task of governments should be to clarify what social, political, and economic objectives they seek. Technologies which are not expensive or large-scale may recommend themselves, but even poor societies seek various benefits in their quest for technology. Access to "modern" technology is sought by all for symbolic as well as for practical reasons. Practical considerations themselves are weighty: technology unlocks new resources, increases productivity, and generates new capacities to produce goods and services. Conflicts arise over goals pursued. Policy-makers in Burma, let us say, or Sri Lanka, may fear that an influx of Western technology will damage local religious, cultural, and family values. Tanzanians in turn may fear that conventional technologies are too heavily biased toward cities, large-scale factories, and control by elitist engineers to fit with their concept of self-reliance. And planners in India may reject Western technologies as too capital-intensive and not sufficiently generative of jobs.

Different calculations apply to diverse sectors of activity. Even a nation committed to "soft" technology may have different criteria for cost, scale, and level of technology in its farming and its mining sectors. It might feel forced to adopt "high" technology in the extractive sector while rejecting capital-intensive technology in industry and agriculture. This suggests that some constraints are inherent in technology: minimal-scale thresholds or the lack of existing alternatives to certain expensive technologies. *The central issue is to choose a whole range of appropriate technologies while clearly defining the purposes they are to serve.*

Many countries wish to optimize their use of local materials, personnel, and financing. One criterion of appropriateness, therefore, is the degree to which technologies foster such optimal use. Those tied to utilization of imported intermediates or to quality control by high-level experts fail to meet this yardstick; they are inappropriate for optimizing local capacities. But what priority scale attaches to such optimization, and how applicable, sector by sector and product by product, is this criterion? If a nation seeks to become competitive in the world market, it may have to sacrifice its desire to optimize local resources. And some double standard necessarily operates regarding quality control. One standard may be suitable for local manufacture, whereas another will apply for export production. Neither safety nor reliability needs be sacrificed, although certain standards of precision, ease of transport, or of packaging can be lowered. This is the criterion applied by the Development Commission of Small Scale Industries of the Indian Ministry of Industry, which has decided that it is better to lose quality, precision, and appearance *for a limited period of time* if these losses are offset by other gains: job-creation, linkages with other producers, and savings in foreign currencies.[27]

The impact of technological choices on job-creation is crucial for policy. Entrepreneurs in many less-developed countries "overautomate" their plants. With an appropriate choice of technology, however, they could generate employment and additional profits. Even when using obsolete machinery makes good economic sense, lack of information as to alternative sources or physical distance from markets militates against the possibility of firm managers purchasing it. (This is true even when prestige considerations do not bias decision-makers against such equipment.) Different criteria exist for determining the appropriateness of technology in the light of job-creation. Much depends on the time span under consideration. In some circumstances more jobs might result, at Period B, from the adoption of more capital-intensive technology at Period A. The point is that technologies are appropriate or inappropriate relative to concrete priorities and time scales.

Much has been written of late on technology and the environment.[28] Third World representatives sometimes portray ecological concerns as a luxury they cannot afford. Some even assert that they welcome pollution if it brings them industrialization and higher material living standards. Notwithstanding such thinking, ecological constraints do bear on the appropriateness of technology. Capital is not the only scarce production factor in less-developed countries; natural resources often constitute another. Hence there is much wisdom in choosing technologies which utilize locally available abundant materials in a nondepleting manner.[29] Choosing technologies according to this criterion generally also facilitates job-creation among unskilled, or traditionally skilled, workers and can result in great foreign-exchange savings. Apart from balance-of-payments considerations, however, such technologies require lower financial outlays in local currencies, particularly in agronomy, health, and construction technologies.[30]

The literature on appropriate technology emphasizes such values as control by people at the grassroots, a human scale of operations, environmental protection, job-creation, and lesser dependency on outsiders. Can any modern technologies be "appropriate" to these development objectives? Although such technologies are appropriate to creating industries which manufacture the kinds of goods produced in societies of origin, the question is whether this production itself is appropriate. (Debate over appropriate technology was, in fact, initiated because of widespread disenchantment with production of this type.) Yet even China, which preaches self-reliance and worker participation in production decisions, uses expensive modern technology in such sectors as communications, electronics, computers, military production, and nuclear development. Locally developed "intermediate" technologies are "inappropriate" in these domains.

But if no single kind of technology is appropriate for achieving all the goals societies set, it does not follow that no qualitative differences exist between "hard" and "soft" technologies.[31] "Hard" technologies are geared to objectives such as power, aggregate growth, mass production, and complex organization. "Soft" technologies, in turn, foster wealth distribution; job-creation; local production; and simplicity in installation, maintenance, and repair.[32] "Hard" technologies tend to bind users to existing dependency systems, while "soft" ones tend to maximize autonomy. Whatever compromises are made in technology policy, sharp criteria must be applied in deciding the mix of technologies deemed essential to follow preferred strategies. If the dominant values sought are ecological integrity and job-creation, departures from this standard will be minimized and contained within institutional limits. China, on this point, seeks to keep its export sector small so as not to produce distortions in its domestic economy.[33] Similarly, urban investment in Cuba and Tanzania are concessions to necessity which do not contradict the dominant policy of "favoring the countryside."

To conclude, an appropriate choice of technologies means choosing a wide variety of technologies, some better suited to one goal than to others. The greatest mistake is to sin at either extreme: to adopt only "high" technology or to rely solely on "soft" technology.

These pages serve as a transition from concern over internal infrastructure to criteria for outside acquisition. They pave the way for a look at Third World negotiating strategies.

Negotiating Strategies

The parties to most technology contracts are firms which sell and purchase know-how. Nevertheless, two invisible parties hover over all negotiations: the respective governments of these firms. Governments of importing firms, especially, increasingly try to structure conditions under which firms within their jurisdiction are to import technology.

It is worth inquiring, therefore, into the issues these governments deem vital in negotiating technology transfers.[34]

Unless negotiating strategy expresses some coherent science and technology policy, it is unlikely to produce desired results: the influx of suitable technologies at reasonable cost, the absorption of technology in ways which optimize local capacities (in financing, personnel, and materials), and minimum dependency. Great variety marks the forms in which governments try to meet these goals. At times, for example, efforts are made to diversify sources of supply, which does not necessarily require "quota" legislation but can be done—in the case of product-embodied technologies—simply by ordering import-licensing authorities to favor certain suppliers. The regulation of process and decisional technologies, however, usually requires explicit legislation prohibiting technology imports when "equivalent" expertise is available from national suppliers and setting payment ceilings for services and licenses. The first clause is usually interpreted flexibly, for although national suppliers exist, they may not be able to supply their expertise at crucial times. Legislation is easier to apply in cases of licensing specific products or processes, rather than in those governing the use of consultants or engineering designers. As for payment ceilings, suppliers can easily find indirect, albeit legal, ways to circumvent restrictions. More importantly, such restrictions do not of themselves motivate national producers to seek out local suppliers. In the absence of alternative infrastructure, the effect of payment ceilings may be contrary to initial intentions. Indeed, until governments can force transnational corporations to submit their accounting procedures to closer checks, ceilings can easily be circumvented. The stimuli and constraints used by governments in negotiation can be seen more clearly if we examine recent proposals of "codes of conduct" for technology transfers.

Disagreement exists over the value of codes and their status. Some plead for exhortatory general guidelines. One example is UN Resolution 2626 (24 October 1970) which establishes an International Development Strategy for the next ten years:

> The Strategy in effect sets up *a code of conduct* for Governments and international organizations. But it does more than reformulate known concepts and give them the status of a world consensus. It contains a number of ideas which could have a profound effect on the future of the economic world order. For example, the consensus calls for a new international division of labor.[35]

Metaphorical codes are also favored by some business firms, for whom codes formalize their beliefs and the standards employees must "live up to."[36] Other advocates, on the contrary, seek a binding international document with the force of a treaty or agreement signed by governments and international corporations, simultaneously regulating duties and rights of each.[37] Drafts proposed by the Pugwash

Conferences, the United Nations Conference on Trade and Development, and a group of Latin American officials who met in Caracas in 1974 are of this type.[38] Two problems plague such efforts: the difficulties of obtaining the agreement of all concerned parties and of assuring compliance in the absence of effective sanctions. Getting signatures is possible, given the pressures generated by widely publicized research on transnational corporations, United Nations support for a code, and the eagerness of business groups like the International Chamber of Commerce, the Council of the Americas, and the Conference Board to arrive at "guidelines for private firms in international trade and investment as well as guidelines for nations participating in the international economy."[39] Most business officials declare, however, that unless they play an important role in drafting codes, their firms will not treat them as binding. This attitude raises the thorny question of compliance, because moral persuasion seems to be the only available international sanction. National sanctions can of course be enforced by governments resolved to do so. But the larger problem is that codes which require the goodwill of signatories are only self-binding, and no institution binds itself to actions which go against its own interests. Furthermore, as one United Nations official writes,

> what the companies themselves, or the business community, regard as responsible behavior may not be looked upon in the same light by society at large.[40]

Corporations could simply make quiet arrangements with pliant governments and mutually ignore the code. This is what international shippers do in the realm of maritime navigation with the profitable compliance of the governments of Panama, Liberia, and Greece.

Notwithstanding its limited value, an international code is still, on balance, worth having for two reasons. The first is theoretical in nature and need only be mentioned here:[41] Normative documents can play a pedagogical, critical, or evaluative role even when power-wielders reject the properly normative function such documents are meant to play. But a second, more practical, reason imposes itself: Considerable momentum has now been generated, and the appearance of some international code seems likely. As draft codes vie for attention and legitimacy, it is essential that some tolerably acceptable formulation be reached even as the fundamental dilemma persists: codes accepted by transnational corporations are likely to be quite harmless, whereas codes with true constraining power are not likely to be obeyed, even if they are signed, by those whose interests they affect adversely. British economist David Robertson contends that many issues of regulation cannot be solved because they require the passing of measures which can only be judged impractical. His reasons are:

- Governments are reluctant to relinquish sovereignty over economic policies.

- Economic benefits from proposed new regulations are uncertain because of our inadequate understanding of the economic role of transnationals.
- Dangers are foreseen in the setting up of quasijudicial committees to assess issues relating to TNCs without an accepted body of rules (and, he adds, this poses endless problems of definition).
- Difficulties are anticipated in enforcing compliance with regulations without destroying the economic benefits accruing from foreign direct investment.
- Fears persist that if a newly established institution revealed itself to be ineffective, the problems created by TNCs might be aggravated.[42]

Robertson favors adopting nonbinding rules and minimum restrictions on TNCs. Jack Behrman, an American scholar, voices similar fears. After noting that draft codes insist disproportionately on measures to curtail undesirable practices in technology transfers, Behrman warns that

> the mere elimination of private restrictive practices (in the hope of creating a free market) will not necessarily prevent the distortion of economic development, prevent social inequities, generate an efficient use of resources, create the socially desirable consumption patterns, or preserve natural resources. On the contrary, all of these goals must be achieved by positive public policy rather than by negative constraints on companies.[43]

What needs to be done, Behrman adds, besides eliminating restrictive practices, is to make sure that all constraints on market mechanisms foster the achievement of the desired objectives. He cites with approval the Japanese approach of defining priorities in types of technology desired and then examining each contract, case by case, to see if it is suitable. "The success of the Japanese policy," he adds, "would argue for a more specific screening process than that implied in a sweeping prohibition of one set of provisions or another."[44] Realism is important if Third World governments concerned over abuses are not to deprive suppliers of technology of the incentive they need to engage in transfers. Such realism may turn out to be one of the more useful by-products to come out of code-drafting exercises. At stake, ultimately, are the criteria for identifying boundaries in negotiation.

Many governments know which harmful practices they must avoid in contracts. The work of Constantine Vaitsos on transfer-pricing points to one arena of friction, namely, monopoly rents disguised as intracompany exchanges.[45] A general concern likewise exists not to allow suppliers to make clients buy a "package" which includes royalties for trademarks and intermediate goods which could be purchased elsewhere. Great resentment arises over efforts made by

TNCs to limit sales to a single market. Causes of friction are well summarized in the recent Mexican law on the transfer of technology:

> Contracts shall not be approved when they refer to technology freely available in the country; when the price or counter-service is out of proportion to the technology acquired or constitutes an unwarranted or excessive burden on the country's economy; when they restrict the research or technological development of the purchaser; when they permit the technology-supplier to interfere in the management of the purchaser company or oblige it to use, on a permanent basis, the personnel appointed by the supplier; when they establish the obligation to purchase inputs from the supplier only or to sell the goods produced by the technology importer exclusively to the supplier company; when they prohibit or restrict the export of goods in a way contrary to the country's interest; when they limit the size of production or impose prices on domestic production or on exports by the purchaser; when they prohibit the use of complementary technology; when they oblige the importer to sign exclusive sales or representation contracts with the supplier company covering the national territory; when they establish excessively long terms of enforcement.[46]

Third World governments can use the sanctioning power of their central banks to block hard-currency remittances outside the country. Brazil and others have further decreed that fees paid to foreign technology suppliers are to be taxed at the higher rate governing profits. Among instruments employed, however—ranging from technology registers or payment ceilings to authorizations for import—none is more strategically potent than the concept of "breaking up the technology package" developed by Andean Pact negotiators.[47]

This phrase is used in two distinct senses. In the first the "package" refers to a cluster of goods and services sold by TNCs to an overseas purchaser. Included in the package are technology proper, permission to use trademarks, equipment, intermediate materials, contractual restrictions on sales to third-country markets, and provisions for supervisory services. Andean legislation dissociates technology proper from the the purchase of intermediates, royalty payments for trademarks (seen as "fictitious" technology), and restrictions on the sale of finished products outside the producing country. A second, more restricted, meaning attaches to the phrase *breaking up the technology package*, this one basic to the struggle of Latin American countries to gain greater technological autonomy. Andean negotiators examine the purely technological components of a contract with a view to disaggregating technology itself into core and peripheral elements. The first step consists of identifying the minimum, indivisible technological modules essential to the process in question. In their totality these elements are designated *modular technology*. Accessory elements are labeled *peripheral technology*. Several criteria are invoked to determine what, in any given operation, is modular or core technology and what is peripheral. To be adjudged *modular* (or *core*), a technology must be, first of all, essential to the process in question. A second criterion for determining this status is the degree to which technological elements are functionally inseparable or, at least, interdependent in a process. Thus, any technology which is indispensable to the success of an entire process or cannot be separated from it is *modular* (*core*); all other elements are *peripheral* (*ancillary*).

The concepts of core and periphery help policy-setters decide how best to lower costs, optimize local inputs, and reduce dependency. Because packages are expensive, local supplying of parts of the package lowers overall costs to purchasers and reduces their dependency on foreign suppliers. More important is the pedagogical value attaching to the attempt to identify core and ancillary technologies. It serves as an apprenticeship which strengthens the hand of negotiators in bargaining, even if they eventually buy parts of the package from outside suppliers. The most difficult task is not breaking up the package but putting it back together again. A proper "fit" of the disparate elements is difficult, and errors abound at first. Nevertheless, the experience gained by Brazil and Argentina suggests that reaggregation of packages is a skill which can be learned quickly. At times outside consultants are hired to help nationals "put the pieces together." And as Gonod notes, the disaggregation of the technical process not only opens up new technological combinations, it also allows its practitioners to free themselves from purely "mimetic transfers."[48] Once the technological myth that there is only "one best way" to proceed is refuted, purchasers notice that hitherto ignored actors in the technological arena possess elements of know-how relevant to their needs. Collaboration between producers and local suppliers creates new attitudes of strength in negotiating with outside suppliers. And local suppliers are stimulated to launch research of their own because they now know that local firms will look to them to supply technology. In order to break up a package successfully, however, negotiators need certain institutional strengths. Initially at least, according to Gonod, state firms able to intervene will have better chances of succeeding than others. A monolithic structure or one in which autonomous and coordinated segments of a process are involved has advantages. On the other hand, if existing structures permit only the coordination of sequential technological processes, package-opening proves more difficult.[49] Breaking the package, in short, requires decisive political will allied to a suitable institutional structure. So true is this that even classifying technology into core and peripheral is largely a function of the vested interest of the analyst.

The Andean Pact has distinguished between core and peripheral technologies in such sectors as copper, nonferrous metals, tropical forest products, and electronics. The informed consensus of their experts presupposes a shared interest in making their nations more autonomous and stronger in bargaining than before. Managers of transnational corporations, on the other hand, or even host-country managers might draw the line differently: their interests lead them to view packages as the most efficient way Third World firms have of importing technology. Most modular technologies are patented and do not fall under categories called by Andean technicians *preprocess, postprocess,* or *general* technology. Predictably, therefore, Andean specialists attack present world patent legislation as biased in favor of corporate suppliers of packages.[50] Since Andean Pact countries are not signatories to the Paris Convention,[51] they have considerable leverage to create new ground rules governing negotiation with foreign investors and suppliers of technology. Yet the sharpest conflicts in negotiation are not over patents but over technological infrastructure itself. "Opening the package" is but one tactic in a larger strategy aimed at shattering the near-monopoly enjoyed by transnational firms in the production of technology. Unless that strategy leads to the creation of viable structures for the national generation of appropriate technologies, it will have but minimal impact. In the judgment of Andean officials, the highest indices of modular technologies are contained in "packages" of relatively new processes and products. The marketability of a package is closely tied to its proximity, in time, to initial production stage. One may thus take the "product cycle theory"[52] one step back and conclude that quasi-monopolistic rewards come to those who lead the innovational "pack." The desire to head this pack has led to the installation of R&D facilities in the first place. Hence, in the absence of its own R&D infrastructure, any country will find its negotiating options curtailed. Even if it limits its imports to "modular" technologies, it can have little control over their price. Power relations cannot be banished from the technical discourse. As Gonod writes,

> technical "discourse" is much closer to economics than it is to science, and it is far closer to social and political discussion than it is to economics itself. Here even less than in other domains can we ignore power relations.[53]

The notion of "modular" or "core" technology can be deceptive because the line between core and periphery is movable. No single element is irrevocably "modular" for, with the passage of time, firm-specific technology can become system-specific or general. New creations can render formerly vital technology no longer essential to a process. Thus the approach aimed at "opening the package" allows "weaker" negotiators to accompany shifts in technology cycles themselves. Ultimately, advanced Third World countries seek to gain control over the direction of these cycles. "Package-opening" is but part of a more ambitious strategy to improve their capabilities as creators of new technology.

Although it predates the contemporary interest of Andean Pact governments in "opening up the package," the following example sheds light on two facets of the strategy they now favor. The first is the need to disaggregate technology into components, some of which can be produced locally. The second is the importance of stimulating local research and development, thanks to appropriate incentives to firms.

During World War II, under the Mutual Aid Program (MAP), the Armed Forces of the United States agreed to supply communications equipment to the Brazilian navy.[54] During the first year the United States would provide to the Brazilian navy 100% of the funds needed to purchase, install, and operate the equipment. By the second year the percentage would drop to 50%; by the third, to 10%, etcetera. Terms were designed to induce the Brazilian navy to spend its own money to keep the project going after launching. Very soon civilian Brazilian suppliers, especially a São Paulo company named Cacique, began manufacturing components for the equipment. Facilitating the competitive entry of Cacique was the general shortage of components caused by the war and affecting US manufacturers. After the war, however, US suppliers no longer suffered from a shortage of materials and were able to displace Cacique from its position as partial supplier to the Brazilian navy. Cacique was eventually bought out by a consortium of transnational corporations, including Motorola Corporation and Telefunken.

At this point the Brazilian navy began applying its technique of "breaking up the technology package" by imposing unique local specifications for its radar equipment and other elements of communications systems. The navy knew that neither Telefunken in Germany nor Motorola in the United States was able to build parts fitting its specifications. As a result, Telefunken's Brazilian subsidiary was obliged to engage in applications research of its own. Subsequently, it designed and built the required parts. In the process the Brazilian subsidiary came to master technology it hitherto had had to import from the German company's main plant.

This example shows that the user of technology can set conditions which encourage the "invention of inventions" by local suppliers. The client in this instance was a governmental firm; but in other cases involving private clients, governments can, through legislation and incentives, create conditions for "breaking up" imported technology packages. The Brazilian navy designed its specifications with a view to making the technology usable as well in a number of "civilian" sectors such as radar for airports. Its objective was to widen the market for Telefunken/Brazil and pressure the company into making the components itself. One policy lesson to be learned is that governments should identify points at which their interests converge with those of TNCs and other actors in the technological arena. They can then harness, by cumulative steps, those convergent interests in pursuit of their own policy goals. Thus it came about that by 1963 the navy and Telefunken/Brazil were fully emancipated from reliance on foreign suppliers.

One parallel measure which recommends itself to Third World countries seeking to reduce dependency is the move to export technology while engaging in selective imports. Establishing two-way flows is an important policy tool.

Two-Way Technology Flows

Most industrial countries acquire some of their technologies from outside sources.[55] Because relative proportions of imported technology are small in comparison with those created by national R&D efforts, little fear exists in industrialized countries, however, that foreigners will dominate national markets or set the pace in vital industries. Foreign acquisitions are viewed with equanimity as valuable complements to endeavors under firm national control. A reverse pattern, however, prevails in Third World sites: most industrial technologies are imported and only a small fraction produced locally. Under these circumstances, importing nations may wish to export some technology so as to set up two-way flows and gain greater familiarity with, and competitiveness in, world technology markets. Becoming an exporter is like gaining entry into the club of influential technology suppliers. Quite apart from its material advantages, the prestige attached to this role certifies one as "mature" in what is generally viewed as a "sophisticated" international arena.

The entry of Third World nations into the charmed circle of technology exporters takes many forms. The simplest is the export of machinery as one form of product-embodied technology. These sales are usually accompanied by service contracts relating to installation, maintenance, and repair. Transactions of this type constitute fairly low-level transfers; greater prestige is attached to exporting process technology, in the form of patents, licenses, and advisory or performance contracts. An even more advanced form consists in providing the expertise needed to conduct feasibility studies, engineering designs, or equipment-specification surveys. All are "decisional" technologies, a category of person-embodied skills, ranging from purely technical know-how to managerial and systems coordination. The provision of these skills by Third World consultants, particularly to clients in rich countries, is highly challenging. Such clients are demanding: they "push" experts to learn in the very process of applying their expertise. This is paradoxical inasmuch as many experts from the rich world regard Third World contracts as more challenging to them because "normal" infrastructure and service networks are absent. Consultants have to improvise more in Third World sites. The point is that technological maturity comes with performing successfully in varied environments. Once maturity is acquired, one can work more creatively even in one's native environment. Firm-specific technologies are particularly enriched by having to be tested in multiple conditions. Some Third World leaders intuitively grasp the dynamics of enrichment and encourage technological exporting. Engaging in two-way flows brings greater knowledge of international demand networks and provides greater flexibility in responding to markets within one's own country.

Some Third World manufacturers export technology to the very firms in the rich world which initially supplied them with know-how. One such case is Hindustan Machine Tools of Bangalore, a public-sector venture manufacturing machines and wrist watches set up in 1953 in collaboration with Machine Tools Buerhle & Co. of Zurich. By 1974 the Indian company had secured an order amounting to four and one-half million Swiss francs through the Swiss partner. Half of this sum was for the export of machine tools, half for technical documentation based on designs supplied by the Swiss company.[56] A different approach was used by Indian Telephone Industries of Bangalore, which collaborated with Britain's Automatic Telephone and Electric Company to make automatic exchanges. The Indian firm has now received orders from its British counterpart to supply equipment which the latter no longer manufactures. A third example involves a private company in Madras, Postons Limited, which has collaborated technologically with the British Associated Engineering Group. The Indian company has now started exporting technology in a joint venture in Malaysia with the Malaysian government. India Postons will supply the entire know-how and a portion of the capital equipment.

Indian production units are viewed by the government as good bases for export to neighboring regions. India's performance in recent years has propelled it to a position of prominence as a supplier of technology to South and Southeast Asia.

Third World manufacturing units often make improvements on technologies supplied to them by a licensor from an advanced country. These improvements can be resold to original licensors. One instance is the sale by a state-owned corporation in Argentina, Fabricaciones Militares, to the Browning company of the United States for improved pistols which Browning had originally licensed Fabricaciones Militares to produce.[57] This practice occurs often enough to generate demands by Latin American negotiators for revisions of licensing agreements, many of which do not allow payments to licensees when these make technological improvements usable by licensors. Were legislation suitably modified, Third World technology exporters would quickly gain access to clients other than original licensors. This is why two-way flows are an important instrument of Third World technology policy.

Two-way technology flows have always accompanied trade in finished products among Third World countries. But such trade was not explicitly designed to enhance technological capabilities of exporting countries, an objective better achieved when process-embodied or decisional technologies, and not product-embodied technologies, are transferred. Competitive position is best gained by competing. Hence countries eager to improve their relative technology positions encourage national technology suppliers to pursue a vigorous policy of internationalization. Corporate managers interviewed in the United States voice the opinion that Third World firms can successfully compete with them in many domains. This is especially true because many trademarks and patents embody no true technological superiority but simply mask what Vaitsos calls "pseudo-transfers of know-how."[58]

Other advantages can accrue from concerted action among Third World actors in technology arenas.

Concerted Action

Concerted action is needed if weak partners in unequal exchanges are to gain bargaining power in purchasing technology. The Third World needs science and technology because the two are "among the crucial tools necessary for increasing national independence and welfare."[59] Neither greater independence nor optimum welfare, however, can be gained by most Third World countries if they persist in acting alone. In 1970 Raúl Prebisch urged upon Latin America a "timely, as well as energetic and enlightened, policy of international co-operation."[60] And Third World solidarity is the keynote of recent meetings held to define the evolving world order. Cooperation in technology means creating horizontal relationships to replace the vertical ones with current suppliers which now prevail, thanks to the near-monopoly industrialized countries have in the generation of technology. One United Nations source estimates that 98% of all industrial research and development conducted outside socialist countries takes place in the developed countries.[61] Given this supply structure, pioneer efforts to promote Third World cooperation stress the need to lower the costs of acquired technology, to facilitate optimal use of local factor resources, and to reduce dependency. Political leaders in Third World countries must perceive technological cooperation as useful, necessary, or indispensable. A graded scale of relative importance attaches to these three terms. Although the lowest degree of attraction is utility, concerted action may also seem necessary once certain goals are postulated. It can be viewed as indispensable if a country's vital objectives cannot be met without it. The Third World quest for a new international economic order may well remain a dead letter unless cooperative efforts are made to "horizontalize" their technological relationships.

Some writers urge establishing regional or international public institutions for industrial research and development.[62] Walter Chudson observes, however, that "this is eminently desirable but likely to be effective mainly or only as an adjunct of significant cooperation in industrial and economic policy generally."[63] Even within the Andean Pact, where broad agreement exists on economic and industrial policy, it has proven difficult to implement technological cooperation.[64] Difficulties are traceable in part to uneven technological levels. Brazil may sell technology to Bolivia and Chile, but it does not thereby gain much useful knowledge for conducting its negotiations with the United States. In addition to mere commercial contracts, institutional cooperation is needed on such issues as information exchange, bargaining strategies, training, regional research and development facilities, efforts to change world legislation on proprietary knowledge, the provision of advisory services to new joint ventures, direct or indirect financing, and the promotion of local values via the development of technologies adapted to these. A brief word on each is in order.

(a) *Information Exchanges.* Better access by weaker technology agents even to nonproprietary information helps break the oligopolistic hold enjoyed by marketing firms from industrialized countries. Poorer countries may find it useful to collaborate in setting up monitoring teams to gather information supportive of the policy of "opening up the package." Thomas Allen points to the critical role played in small and medium enterprises by "technological gatekeepers" who serve as liaisons between users and suppliers of technology.[65] His study suggests that many problems faced by users of technology could be solved easily if they knew more about available technologies. Jan Tinbergen, thinking in a parallel vein, advocates creating a "new autonomous institute for technological exchange to link suppliers (enterprises and universities) with user-countries. . . . In this universities and user-countries could act as counter-weights against enterprises."[66] That such exchanges are important to Third World users is evidenced by the continuing demand made on providers of technical answering services. The goal of cooperative information exchanges is to facilitate systematic transnational technological gatekeeping.

(b) *Bargaining Position.* Spokesmen for transnational corporations advocate pooling information to strengthen their bargaining stance with the Third World.[67] Parallel efforts by Third World representatives center on adopting common positions on such issues as economic rights, sovereignty over resources, and controlling activities of foreign suppliers of technology. Considerable Third World pressure has been applied to gain approval of codes of conduct governing foreign investment and technology transfer.[68] In large part, these codes bear on norms for negotiated technology purchases, training national personnel, rules for technical supervision *in loco*, and terms of sales to third countries. Were large numbers of Third World countries to reach a common position on these issues, they could negotiate more equitably with transnational enterprises. Success in codes, however, is conditioned by the degree of political commitment governments assume to enforce rules.

(c) *Joint Training.* Argentina has pioneered an autonomous approach to nuclear-energy production using natural uranium.[69] Meanwhile, India has gained considerable expertise in small industries; Algeria, in petroleum and natural gas. These examples hint at increasing Third World abilities to provide training in some sectors, which is another way of "horizontalizing" relations. A major obstacle is the abiding fascination of many Third World professionals with diplomas from prestigious rich-world universities. No easy way exists for breaking the stranglehold this imagery has on them. Yet no serious gains in technical maturity can ensue until new standards of status attribution are translated into training actions. Unfortunately, many Third World policy-makers live in a dreamworld characterized by unrealistic dualism. On the one hand, they criticize rich-world training institutions for preparing Third World professionals in ways unsuited to their local responsibilities. On the other, they refuse to send their own trainees to other Third World sites, invoking as their rationale the fear of "second-rate" training and apparently internalizing the myth that institutions in the "developed" world are intrinsically superior. At some point, Third World policy-makers must simply break with the existing prestige system and take steps toward eventual technological autonomy. The road to autonomy lies in building a network embracing training policy, professional incentive systems, and criteria for weighting the relative claims of efficiency and lessened dependence. Unless they take bold steps to create autonomy, Third World nations will continue to reap a harvest of inapplicable technologies.

The establishment of joint Third World technology training institutes in selected sites is highly desirable.

(d) *Cooperation in R&D Infrastructure.* The usual argument against building R&D facilities in LDCs is that domestic markets are too small to support them. But the argument is largely spurious, given that there are diseconomies as well as economies attaching to large scale. The success of small, high-technology firms in the United States suggests that loss of flexibility and the inability to make rapid responses are such diseconomies. And whatever be the merits of standard arguments on scale, regional Third World efforts seem warranted.[70] Sagasti and Guerrero would like to mobilize Andean subregional talents to create their own transnational corporations, some of which ought to engage in research and development on behalf of the region.[71] Larger countries like India, Brazil, and Argentina can identify specific sectors in which national R&D is warranted in scale terms.[72] Imaginative policy might lead them to assign a regional role even to national R&D installations along lines of the "leased-time" concept applied to computers in the commercial world. As for smaller or poorer countries, no *prima facie* obstacle stands in the way of OPEC or other Third World associations subsidizing R&D installations geared to their needs, at least in a few vital industrial sectors.[73] Many research directors in US-based firms whom I have interviewed declare that the main obstacles impeding such novel endeavors are not financial, organizational, or even technical but political in nature. Roy A. Matthews, research director of the Canadian Economic Policy Committee, rejects this view, however, arguing that "probably no other arrangement [he is speaking of the system of developing technology at the center, that is, in the developed countries] could so effectively permit home-country industry to have access to the latest technological advances and disseminate their benefits in the form of economic enrichment to the population." Matthews acknowledges that his arguments have little hold over policy-makers in the Third World, because "the issue here is simply not an economic one: it is a matter of cultural affirmation."[74] The Third World aspires after its own R&D capacity.

(e) *Changing World Legislation on Proprietary Knowledge.* There is no inherent reason why technological applications are restricted to their investors or institutional owners. Moreover, the protection of industrial property has never been an end in itself but a means to encourage industrialization, investment, honest trade, more safety and comfort, less poverty, and more beauty in the lives of human beings.[75] The basis for considering knowledge to be proprietary is positive law, expressed in the form of international agreements and national legislation on intellectual property. The rationale for legal recognition of monopoly rewards given to inventions has always been that in their absence invention and improvement would not be adequately stimulated. Yet, as one student of technology writes, "there is an over-emphasis on the licensing of new sophisticated inventions; most of the manufacture done in the world uses either non-patented processes or processes on which patents have run out."[76] Perhaps so, but the "competitive edge" enjoyed by large transnational corporations derives mainly from their proprietary technology.

The long-term effect of a legal system which creates economic monopolies for creators of technology is to perpetuate the advantages of those already favored by present structures. Thus if current rules governing proprietary knowledge are maintained, poor countries can never achieve relative technological parity with the rich. Unless technology becomes "the common patrimony of the human race," inequitable rewards will continue to be assigned to those who already enjoy a privileged "competitive edge" in technological arenas.

The long-term task consists in creating a noncommercial basis for technology-sharing on the basis of priority need. Progress will be slow, but Third World countries must begin to concert their efforts with a view to revising industrial-property laws. One objective, of course, is to alter the role of research and development activities as mere adjuncts of profit-seeking enterprises. Eventually it may become necessary that R&D be made into permanent activities conducted by a "world community" which takes as its first priorities worldwide social justice, material sufficiency for all, and ecological integrity. Enlightened corporations profess concern for these values,[77] it is true, but only on condition that they also profit by them. But profit-seeking itself must be subordinated to wider values and become an instrumentality of resource allocation, not its organizing principle. National legislation in several LDC countries already rejects trademarks as legitimate proprietary knowledge. This is a step in the right direction; its logic should extend to many technologies, perhaps eventually to all knowledge considered proprietary.

(f) *Constraints.* Obviously many Third World governments will shy away from cooperation with others, lest they themselves lose out on major benefits in the highly competitive technology arena. Mutual distrust is the major political obstacle blocking concerted action. Yet this obstacle is not absolute; Third World solidarity can be reinforced to overcome suspicion and other constraints: claims of existing international organization, the inertia of national bureaucracies, and diverse levels of technological development in the Third World itself. Levels of development are dynamic, however, not static. The technological capabilities of less-developed countries, as Raymond Vernon points out, change rapidly over time.[78]

The lesson to be drawn is that technological complementarities can exist among Third World nations even under conditions of great initial diversity in technical levels. These complementarities can be created.

Concerted Third World action in technology is difficult, but chances of success are heightened if national efforts are supported by suitable policies in international agencies.

International Support for Third World Technology Policies

Great importance attaches to the *mode* and the *conditions* under which access to technology is gained. Do these favor such values as a better material life for all, modern and efficient institutions, greater social justice, enhanced opportunities, and ecological integrity? Any international support must contribute to these goals. The general assumption here is that international organizations are free to undertake actions consonant with Third World desires and not normally possible for other international actors. This has been true in technical assistance, development-planning, and financing. Embryonic steps are now being taken to expand the scope of such "support" to technology policy.

As a start the United Nations has created a Working Group on Technical Co-operation Among Developing Countries with a mandate to recommend ways for developing countries to share their experience with one another so as to improve development assistance and to investigate possibilities of regional and interregional technical cooperation among developing countries.[79] Although the major emphasis to date has been placed on inter-Third World fellowship and training programs, greater attention is now being given by the Group to narrowing the "communication and information gap as regards the technological capacities of developing countries."[80] There is no need to review here the recommendations of the Group. The important point is that cooperation has been accepted as a valid principle and that mechanisms to implement it have been set up. Poor countries can gain better access to the range of technologies by the international provision of a variety of services: information catalogs and rosters, answering services supplied by various agencies, and mobile quality-control teams to help Third World industries decrease their reliance on expensive technological consultants from private firms in the rich world.[81] Supportive roles need not stop here, however.

International organizations have provided subsidies to LDC consultants to gain contracts in other Third World sites in order to lower technological costs to Third World acquirers and reduce their dependency on rich-world consultants. These activities are exercises in collective self-reliance. The Organization of American States' interesting Pilot Project in Technology Transfer lays great stress on developing national and regional focal points which serve as centralizers of technological demand and elicitors of supply.[82] Among lessons learned from the experiment is a general sense that alternatives to technological dependency are possible once facile optimism and fatalistic resignation are purged. Some imagination is required to visualize forms of international activity congenial to sound Third World national, regional, and interregional technology policies. UNIDO, or some similar organization, might stage an international fair to exhibit "appropriate" technologies, confining exhibits to technologies developed within the Third World which hold the promise of application in other sites. Multiplier effects could come by obtaining international subsidies for the dissemination of technologies successfully tested. Such a step would help create an alternative to the present expensive single-channel technology marketplace.[83]

The United Nations or some other international entity might also charter a capital-replenishing (if not profit-seeking) firm to play functions of research without which fairs will have no abiding impact. Initial steps in this direction need not be tried on a fully international scale; collaborative regional and local efforts can proceed on similar lines. Eventually, coordinated planning will be required. One corporate spokesman thinks that such broad planning should include adversaries of LDCs, that is, transnational corporations. For André van Dam "the heart of the matter is . . . the underlying concerted planning, research and development between transnationals, and between transnationals and home and host governments."[84] He favors creating a transnational research and development institute for making breakthroughs in appropriate technology and for "seeking out the full range of development tasks where the priorities of the Third World mesh with the unique resource capabilities of the transnationals."[85] Van Dam's view illustrates the need increasingly perceived in governments, international agencies, corporations, universities, and think tanks for comprehensive collaborative planning aimed at finding systemic solutions to problems. International agencies can play an important role by virtue of their special legitimacy, at least latent, for speaking on behalf of "all" interests and not primarily for limited vested interests. But they can do so only if they alter their present modes of operating.[86] The point is simply that the creation of national technology policies consonant with genuine development depends upon supportive actions from international agencies for success.

* * *

The present chapter has identified the objectives and general directions of Third World technology policies. Most countries seek to gain access to the whole gamut of modern technologies, at a fair price, in a mode which allows them to make optimal use of their local resources and in ways which minimize their dependency on outside suppliers. In order to translate general desires into concrete policy, however, any society must assess its constraints and the effective leverage it has for implementing its wishes. Both constraints and leverage vary widely. Large countries like India, Brazil, Argentina, or Algeria may seek a degree of technological "modernity"—even by following the path of massive technological imports—while sacrificing on other objectives (for example, optimization of local resource use) because they hope, eventually, to achieve relative technological sufficiency. Many smaller countries, on the other hand, may find that their chronic need to purchase foreign technology invalidates any hope of gaining relative technological autonomy. Beyond such variables the decisive elements in policy are the basic options and development strategies chosen by a society. The "vital nexus" which links options, strategy, and policy to one another powerfully influences the approach to technology even where governments pay little overt attention to technology policy.

No single country can fully satisfy its desires in technology matters, if only because of inherent tensions between the quest for greater autonomy and the wish to gain access to all "vanguard" technological flows. The objective of keeping technology payments down may likewise chase away certain suppliers. What is important is to relate science and technology policies directly to overall planning objectives and to assure that measures taken to attract or control technology be internally coherent and supportive of society's broader goals. Individual countries need to discover new combinations of sound technology policy, allied to horizontal collaboration with other Third world actors, and support by appropriate international actions.

The instruments used by Third World governments—mandatory registry of technology contracts, ceilings on payments, requirements to use local technology where available, the break-up of negotiating packages, and the like—tend to reduce dependency and lower costs of acquired technology. Control alone is not enough, however; national and regional science and technology capacities must be built up. Hence the emphasis on new forms of training, creating indigenous R&D facilities, initiating two-way technology flows, and promoting a range of "appropriate" technologies responsive to special factor endowments and larger social objectives. One key to success is the Sabato "triangle" which links—institutionally, informationally, and functionally—the demand, supply, and policy actors in the technology arena. Internal integration must occur in every vital sector of the domestic economy, as well as with imported technology. Such integration is dictated by the need to assure national assimilation, mastery, and dissemination of acquired technique. Throughout all actions, incentives are central—to purchasers of technology, to intermediate agencies (financial, consulting firms), and to the actual and future human resource pool. Supportive action from international agencies is also important. Although no mention has been made in these pages of assistance from the rich world and its agencies, public and private, this is also vital, as evidenced by the seriousness with which Third World leaders promote the drafting of binding "codes of conduct" for investment and technology transfers. If realism is to prevail, inner consistency and broader systemic congeniality between technology policy and overall social goals must be based on a sober assessment of existing constraints, allied to deliberate choices as to timespans over which limited targets will be pursued. Because it is never possible to make great gains without paying social and human costs, the criteria for deciding which costs will be viewed as tolerable must be defined, preferably after consultation with the intended beneficiaries of policy and the populace which must bear those costs.

Third World policy-making does not take place in a vacuum. The context of LDC technology policies is an international order dominated by transnational corporations, international agencies, and big-power governments. This order is not static, however, but is undergoing rapid change. What has been said thus far on the nature of technology, the channels for its transfer, and preferred Third World technology policies must be seen in its global matrix—an evolving world order.

9/Technology Transfers in Context: An Evolving World Order

Writings on the international economic order now proliferate.[1] This "order" comprises all the networks which channel and regulate international exchanges of money, information, goods, services, equipment, and personnel. National and subnational economic systems are inextricably caught up in webs of interdependence affecting trade, investment, monetary systems, foreign debt, aid, environmental issues, the ocean regime, science and technology, development models, LDC negotiating strategies, and the public understanding of development issues in rich countries.[2] No less important than global economics are the international legal and political orders which link diverse actors—governments, international bureaucracies, private organizations, foundations, professional groups, and churches—around problems of power, legitimacy, and exchange.[3] Indeed, although they are distinguished for purposes of analysis, the economic, legal, and political orders are in reality but aspects of a single "global system" within which technology circulates and makes its impact on development. Influential actors in the technology arena are *ipso facto* important agents within the global order. Conversely, institutions vital to any of the three interlocking international orders are, by that fact alone, deeply involved in the universe of technology.

The world system and the technological universe are both undergoing rapid evolution. Consequently, technology transfer must be seen in the context of the dynamics of the present world order, the forces shaping its change processes, competing images of the future order, and issues on which world-order questions directly affect development and technology. Most writings on development use the term *international economic order* (IEO) generically to embrace the economic, legal, and geopolitical global systems. It is helpful to analyze briefly how the IEO, in this generic sense, affects Third World development.

Impact of the IEO on Development

The workings of the IEO affect development efforts of poorer countries at several vital points. The first is the link between the IEO and social injustice in many nations. The present order tends to freeze the unjust division of the world's wealth in a manner which favors privileged classes within LDCs while constituting a global privilege system in its own right. The current economic order was designed by rich countries to serve their financial and economic needs. This is why it fails, by and large, to meet the needs of less-developed societies, notwithstanding "aid" programs which transfer a tiny percentage (often less than 1% of GNP) of financial and technical resources from rich to poor countries.[4] Exchanges within the IEO respond to purchasing power expressed in competitive markets. The principle of buying power inherently favors rich nations, classes, and interest groups to the detriment of the poor who, by definition, lack buying power. Much "aid" reinforces market exchanges by subsidizing consultants, technicians, and administrators who act as intermediaries between "donor" agencies and "recipient" countries. The IEO derives its legitimacy from a conceptual superstructure elaborated by economists as "laws" of international exchange. One key element in the system, the theory of international trade, had as its purpose, according to Myrdal, "*the explaining away of the international equity problem.*"[5] Development scholars are increasingly concerned with the impact of the IEO on national policies for several reasons. First, the growing knowledge possessed by the "international community" regarding China's performance has widened the stock of development paradigms. This is relevant because no nation preaches so loudly as does China the merits of shutting out "nefarious outside influences." In addition, the failure of import-substitution policies widely championed in the 1960s and the general ineffectualness of national planning cannot be explained unless one assesses the effects of the IEO on national development efforts.

A second realm in which the IEO touches national development is the relation between internal privilege sytems and outside dependence. Underdevelopment can best be understood relationally: privilege systems in poor countries find their normal reinforcement in alliances between national elites and international investors, traders, and professionals. The prevailing IEO sets the patterns of decision-making within which underdeveloped nations relate to industrialized nations and to each other. Shifts in the relative power of the rich and the poor world now challenge the old order, but these changes benefit mainly a few "newly rich" nations in the Third World. In no fundamental way do they alter the competitive ground rules of global exchange.[6] They simply acknowledge the thicker bankbooks of new actors on the international monetary scene. The IEO is attacked more basically by the "dependency" theorists who contend that global networks of investment, trade, financing, aid, technology transfer, and the marketing of "consumerism" foster exploitation within Third World societies. The emphasis noted at gatherings of the UN Conference on Trade and Development, the General Agreement on Tariffs and Trade, and the United Nations is different, however. Here attention focuses on tne disproportional (and, by implication, the unjust) bargaining power of rich and poor actors in the international system. Earlier, the Pearson Report had spoken of "partnership" in development and of the "global village" but studiously avoided mentioning that this village is ruled by village elders—developed nations and their allied interest groups. Tibor Mende comes closer to the harsh truth when he concludes that even altruistic "aid" is largely an exploitative device.[7] Interdependence is rife in the global village, but little reciprocity can be found therein. To offset this lack of reciprocity, many Third World countries pool their votes, their rhetoric, and in some instances even their resources.

Without reciprocity no international order can foster authentic development for all. Genuine development is the symbiotic combination of certain tangible benefits (the *what* of the development process) and humanizing modes in which these benefits are sought (the *how* of the process). If only one ingredient is present, there is no genuine development. It is not enough to improve material conditions, modernize institutions, or achieve self-sustained growth. All these benefits can be obtained in a counterdevelopmental mode: dictatorially, paternalistically, or in unjust patterns of distribution. Conversely, an exclusive emphasis on modal values alone—participation, egalitarianism—may also prove counterdevelopmental, leading to inefficiency and/or stagnant or parasitical employment policies. Substance and style (the what *and* the how) are equally important and must be pursued in tandem, even though a creative tension pulls them apart. As applied to the IEO, this principle of complementarity means that something *qualitatively* other than a mere redistribution of resources among nations is required, specifically, genuine respect and a voice for all in framing ground rules governing international exchange. Powerless or poor nations are not to be treated as global "charity cases," but an international order without reciprocity assures that poor nations will be so treated, even if elaborate disguises or purely nominal concessions in procedure or rhetoric veil inequalities. Reciprocity in the IEO dictates that the poor and powerless, like the wealthy and influential, enjoy access to essential world resources *upstream and not merely downstream.*[8] A new social charter defining the basis of initial claims on resources must deal with initial access to resources, not merely with subsequent distribution. I shall return to this notion later in this chapter, but it suffices here to note the relevance of the concept to the very possibility of development in the Third World. In its simplest form, the argument states that an international economic order controlled by a few rich nations—whether exclusively "old rich" or "generously" inclusive of the "new rich"—cannot be an equitable and a just developmental order. This is true even if extra "special drawing rights" are assigned to Third World nations, seabed funds created for their investment needs, or international subsidies given to their technological research activities. All nations must not only have legitimate access to vital resources; they must also have an effective voice in decisions governing the use of these. Otherwise, development becomes a mask for paternalism or elitist social control of needy masses.

Two points stand out: (a) technology transfers take place against the backdrop of an international economic order; and (b) the evolution of the IEO bears directly on the Third World's quest for more justice, autonomy, and reciprocity. Before inquiring into that evolution, however, we must establish the identity, values, and interests of the major institutional actors in the present international economic order. Predictably, these actors are also those who play a dominant role in the global circulation of technology. I say "predictably" because technology exchanges are a reflection of the relative power and the values operative in the global order at large.

Institutional Actors in the IEO

The global stage is peopled by a rich cast of actors who serve as institutional links between specialized economic, educational, or informational activities and international political relations. The players include international labor unions, religious and missionary bodies, cultural-exchange societies, recreational clubs, and others. But for present purposes it suffices to look briefly into the role and values of five actors: national governments, public international agencies, transnational corporations, world knowledge specialists, and the world communication system. Taken together, these groups decisively affect world systems: they are, in effect, the stewards of technology policy and an elite world coalition.

National Governments

National governments are important actors in the IEO, notwithstanding the laments smaller ones make over their relative weakness in the face of corporate giants. A qualitative difference doubtless exists between the influence wielded by small and medium powers and that of big powers. Yet even small nations can play significant, albeit limited, roles in the global arena. Their legal sovereignty gives them international status and instant access to forums not open to other institutions. Only states can have diplomatic representation and worldwide recognition of the "legitimacy" of their use of armed force. Mere existence as a sovereign state *ipso facto* confers entry to the world stage as an actor. Nevertheless, within this category great powers and a few middle powers are dominant for several reasons.

Rich countries serve as home bases whence investment and technology are exported to the Third World. They are also the sites where consumer paradigms having great suggestive power worldwide are created. Finally, although rich countries rarely act in full concert, they do define the military and geopolitical rules for survival in the world polity. Accordingly, the self-images big powers have constitute crucial variables for development possibilities in the world. The United States and the Soviet Union still hold a privileged place in the galaxy of governmental actors. And it matters relatively little whether the management of the global order is shifting from a two-pole to a five-pole model;[9] in both models the definitions these two superpowers make of their national interest and security decisively affect the workings of the global order. The United States has long adhered to the imagery of "balance of power" and "spheres of influence" as foundations of its foreign policy. Under Henry Kissinger's tutelage recent administrations have given new popularity to the system.[10] Yet the balance-of-power approach to the world by a big power relegates Third World concerns to the periphery. Richard Falk writes of Kissinger's approach to the Third World that

> it sustains the rich and powerful, while it exploits and pacifies the poor and weak. It chooses a globalist organization based on hierarchy rather than equity.[11]

The assumptions underlying balance of power and spheres of influence are radically incompatible with a world order fully congenial to genuine development for all—large and small, rich and poor. If the global order is manipulated by great powers—and, again, it matters little whether the club has two, five, or six members!—their intervention in troublesome areas is legitimated. This model is but an updated version of the civilizing mission of "advanced" countries, bearers of modern technology who bring developmental redemption to "backward" lands. This scheme renders true reciprocity in world exchanges, respect for national diversity, and effective Third World participation in global decisions impossible. Worse still, excessive focusing on balance of power assigns to the self-defined national-security interests of great powers a disproportionate weight in world decisions. Resistance to these ideas has already begun, however.[12] Many Third World leaders repudiate this vision and defend new forms of solidarity as a necessary means for assuring participation by weaker nations in world decisions. Champions of this newer view grant the merits of East/West *détente*, big-power security, and intergovernmental cooperation, but they insist that these goals are not to be gained at the expense of the Third World. Of course, Third World critics of the great powers are not themselves without fault; they are often, in the words of Fouad Ajami, "unduly nationalistic and parochial."[13] Nevertheless, they correctly stress the dangers of perpetuating big-power monopoly over the management of world economic, legal, political, and cultural affairs. They thereby help to expose the basic interests and values of big-power governments.

International Agencies

International agencies range from the World Bank and regional development banks to the United Nations, its specialized agencies, the Organisation for Economic Co-operation and Development, and assorted world-relief agencies. These institutions have in common international membership (if not always leadership or financing), mandate, areas of operation, and overt value perspectives. Most of them engage in some activities centrally related to Third World development. Thanks in part to the self-assigned role played by Robert McNamara, the World Bank aspires to serve as development pedagogue to the rest of the world. McNamara lets no major world conference pass without updating his prescriptions to the world "development community" (in the form of analytic exhortations incorporated into his formal speeches) and calling the attention of specialists and citizenry at large to issues such as unemployment, income distribution, or assistance to small farmers. Not surprisingly, therefore, a 1962 report issued by the Hazen Foundation considers that "the World Bank and its activities represent a significant expression of a new type of international sovereignty."[14] Seen in this light, the discussions which followed the publication of development reports in 1970 (Pearson, Peterson, Tinbergen, and others) take on new life. At that time many students of development asked whether assistance to less-developed countries was better given bilaterally or multilaterally. Although wide agreement favored the latter, some Third World leaders point out privately that at least they have at their disposal means for creating counterweights to governments supplying bilateral aid. They can apply diplomatic and political pressure on them, nationalize or threaten assets of their nationals, or allow their public opinion to mount embarrassing publicity campaigns. But, they ask, how can they express their dissatisfaction with the World Bank, the United Nations Development Programme, or some international funding agency? Such entities relate to Third World host countries solely as "donors," and no other arenas exist justifying their presence there in other capacities—as political actors, let us say, or as holders of economic interests. Not surprisingly, some writers criticize the "leverage" used by the World Bank to interfere in development strategies of Third World nations.[15] Defenders of the Bank reply, not implausibly, that "a bank is a bank is a bank" which must look to "credit-worthiness" more than to "worthiness" as defined by the simple criterion of mass need. Because the Bank must itself obtain capital in competitive markets, it claims it is justified in imposing professional standards of reimbursability. Bank officials point as well to other "windows" in their lending counter—the International Development Association where "soft" loans with high grant components are made on quasiconcessionary terms—and the "third window" for balance-of-payments assistance tied to inflationary prices. Whatever be the merits of this debate, the influence of international financial agencies on the world developmental stage is undeniable. To a great extent, these agencies set the terms of global debates on development. With a view to countering the terminological and conceptual dominance exercised by such groups, a number of Third World intellectuals meeting in Santiago, Chile, in April 1973 created the Third World Forum.[16] The very creation of the Forum attests to the importance of international institutions in shaping the language of development in ways congenial to rich-world interests.[17]

International development agencies fund many consultancy activities bearing on technology transfer. An earlier chapter describes the role of such firms in prefeasibility and feasibility studies, in diagnosing development problems and choosing strategies, and in evaluating programs and projects. Numerous contracts given to rich-world consultants to work in the Third World are possible only because they are subsidized by international institutions or bilateral aid agencies. And if indeed the ability to conduct feasibility studies with local resources is the best touchstone for judging a Third World nation's technological maturity, international funding agencies are evidently major actors in the world order. They reinforce prevailing standards of competitiveness in such crucial domains as diagnostic and prescriptive technologies. By their involvement in this role, international agencies ally themselves closely with the interests and working styles of another actor on the IEO stage: transnational corporations.

Transnational Enterprises

Transnational enterprises have now been exposed to the glare of worldwide publicity. Spectacular abuses and political bribes account only in part for this publicity. It is the rapid spread of public knowledge as to the size, power, and bargaining position of TNCs in the Third World which explains the attention now showered upon them by universities, legislators, international task forces, private scholars, and church commissions.[18] It is superfluous to repeat here what is written above on the role of TNCs in technology transfer. Nor is this the place to review the findings of such authors as Vernon, Levitt, Dunning, Kindleberger, Perlmutter, Girvan, Barnet, Müller, and Turner. What is useful is to summarize the general assumptions and value preferences adopted by different categories of research on TNCs. Such research is no less important a reflection of how TNCs perceive their critics than the "concessions" companies would accept in any transformed IEO. The president of Business International characteristically pleads for

> rules of the game that are clearly stated, harmonized, and to which the international corporations could conform. The reason no code or unified rules of the game exist is not because these are unacceptable to the international corporation (most international corporations would welcome them), but because nation-states have been unable or unwilling to yield sufficient sovereignty to make possible the framing of such rules, or such a code of conduct.[19]

Sovereignty is undoubtedly the most conflictual issue opposing TNCs and governments of host and home countries. At issue is a redefinition of the basis for sovereignty. Its traditional foundation has been legal recognition of a society's political organization as a nation-state. Yet shifting realities are leading some to urge attenuations of political sovereignty—or the conferral of economic sovereignty—on other grounds. One Latin American dependency theorist cites with approval the following passage:

> The international corporation is acting and planning in terms that are far in advance of the political concepts of the nation-state. As the Renaissance of the fifteenth century brought an end to feudalism, aristocracy and the dominant role of the Church, the twentieth-century Renaissance is bringing an end to middle-class society and the dominance of the nation-state. The heart of the new power structure is the international organization and the technocrats who guide it. Power is shifting away from the nation-state to international institutions, public and private. Within a generation about 400 to 500 international corporations will own about two-thirds of the fixed assets of the world.[20]

Some suggest that TNCs be given *de jure* voting rights in the United Nations, in recognition of their *de facto* sovereignty in many weak lands.[21] Others publicly wonder whether selected firms might not be internationally chartered in some Caribbean isle acting as a Vatican State for business, far from the jurisdiction of any government, and perhaps even be admitted to membership in the United Nations.[22] Others wonder whether corporate personnel should not be granted international citizenship so as to facilitate still further their mobility across "purely national" (and, by implication, artificial and arbitrary) boundaries. The most disturbing idea implies that most Third World states are not administratively, economically, and politically viable. Therefore, their leaders might be led to consider contracting the running of their entire countries to transnational corporations, because these alone possess the resources, skills, personnel, and experience to make those states "work."[23] Most researchers judge transnational enterprises, on balance, to be indispensable and beneficial to the Third World. They believe that no other institution can play their multiple roles: raising large sums of project capital; recruiting skilled teams needed to do feasibility, site, and design studies; "transferring" technologies, preferably as packages;[24] and responding flexibly to new "market opportunities." Raymond Vernon declares that TNCs continue to be welcomed because they undeniably bring capital, technology, jobs, and products to numerous sites where these were previously lacking. For him the most important truth is that no realistic alternative to TNCs lies in sight. Vernon argues that "normative" questions (such as, What is right? How ought enterprises behave?) cannot be discussed rationally until "all the evidence is in." He insists, therefore, that the task of "serious scholarship" is to find out all the facts, more specifically, to study how TNCs operate in diverse Third World environments, what their pricing and wage policies are, and what impact they have had on different societies.[25] Vernon's opinion is important for two reasons: He is the symbolic leader of an influential "school of thought" at the Harvard Business School, and his views rest on vast stores of empirical information. Most global corporate spokesmen are sympathetic to Vernon's research because he treats categories such as dependency, exploitation, and control as "external" to arguments about TNCs. And thanks to refinements in his "product cycle" theory, Vernon can assert to Third World interlocutors that oligopolistic advantages accrue to companies only for a limited time and that tougher host-country bargaining stances generally lead corporations to make flexible accommodations. The key, says Vernon, is for LDC governments to be likewise flexible so that TNCs will not be chased away by excessive restrictions. Firms want stability, but this does not mean that they are reactionary or that they favor dictators. They fear abrupt change, whether to the right or left, because such change disrupts their efficiency and impedes sound corporate planning.

Barnet and Müller, Pierre Judet and Jacques Perrin, Norman Girvan, Kari Levitt, and others retort that the Vernon position is, ultimately, little more than a scholarly rationalization of corporate values. They claim that issues of social justice, people's participation, lessened dependency, and the priorities of national development should take precedence over corporate interests and lead to new ground rules. Methodologically, they add, one must study the problem in its total patterns of political economy, not in piecemeal positivistic fashion. Questions of power, control, ideological conflict, and elite decisions are not "externalities" but essential factors in appraising TNCs. Moreover, these scholars (and others, such as Stephen Hymer and Laurence Birns) declare that the evidence reveals that TNCs have indulged generally and systematically in exploitative practices, including exorbitant transfer-pricing, the imposition of unsuitable technology packages, denationalization of Third World capital, encouragement of the "brain drain" of national skilled people to the international market, and the abuse of tax shelters in discriminatory fashion. Notwithstanding the harshness of their criticism, however, these writers do not advocate destroying TNCs but instead suggest methods to minimize the damage they can do. Their recommendations center on greater public disclosure of financial data and on specific measures that would allow bargaining partners to gain countervailing power in their dealings with large firms. This conclusion is also reached by other researchers on transnationals: UN agencies and teams, church groups, individual scholars, and special-interest groups such as labor unions, national governmental commissions, and radical political movements.

Corporate organizations react, in the main, by reasserting the legitimacy of profit-making in a socially responsible manner and by branding many accusations as unfounded. They hold that most "abuses" represent exceptional aberrations from mainstream corporate behavior. More importantly, they maintain that TNCs should not be faulted for *not* behaving as philanthropic foundations, charitable missions, or even as developmental planners for society at large. Corporations reaffirm their loyalties to all societies (or to none!), to all ideologies (or to none!), and to all social systems (or to none!). Finally, they are expending great effort to convince the world that any legitimate and reasonable demands made by the Third World are fully compatible with mature, professional, partnership relationships with TNCs.

As debate, polemics, and research continue, every hue of opinion can be found. The only point of which all parties agree is that transnational corporations are vitally important actors in the world's development arenas.

World Knowledge Specialists

World knowledge specialists constitute the fourth galaxy of stars in the international order. Taken together they consist of a loose, at times barely visible, consortium of universities, scholars, foundations, research institutes, assorted "think tanks," and international federations of study institutes.[26] Their loyalties are global, as are their arenas of action, patterns of expenditure, and travel habits. This cosmopolitan flavor is most evident in the "scientific community." Diana Crane writes that "basic science is an inherently international activity. Its principal goal is the production of new knowledge which is evaluated according to universal standards. In terms of membership and goals scientific communities have been international since their emergence during the seventeenth century."[27] Later arrivals as members of the "international club" include social scientists, historians, philosophers, systems analysts, and the new breed of "futurologists." Significantly, however, most "intellectual powerhouses" are located in, or funded by, metropolitan "developed" countries. One major role played by the "international intellectual community" parallels that of literary academies within national societies: to guard the purity of development language from contamination by indigenous Third World upstarts. Words, concepts, images, theories, and models must be certified as legitimate by the international "intellectual" community. When submissive acceptability is not forthcoming, however, the world knowledge industry subtly proceeds to disarm new ideas, new terminologies, and new models so as to incorporate them into mainstream thinking. The objective, of course, is to take the sharp bite out of *dependencia* theory, theology of liberation, revolutionary *conscientização*, and other efforts by Third World intellectuals to define their own reality as a prelude to prescribing change.

Although this world knowledge consortium is loose and not always fully visible, this does not mean that universities have no walls or that research institutes are staffed by angelic spirits, much less that their assets are other-worldly. What is meant is that scholars, educational institutions, think tanks, and foundations do not automatically qualify as members of this international knowledge "jet set" simply by existing. Throughout academia, research institutes, and foundations can be found certain niches where individuals or teams use their organizations as bases for "keeping in touch" with international conferences, seminars, workshops, ideas, peers, loyalties, and new funding opportunities. Taken collectively, these people and groups gain an "inside track" along with other powerful actors on the world scene: international agencies, transnational business, governments, and a host of "public interest" groups each having its "private" agenda. The important point is that their ideas are listened to by influential decision-makers. In addition, members in good standing of the international intellectual club recruit and screen new members of transnational professional associations in science, the social disciplines, and the multidisciplinary studies.

The world knowledge system plays two roles: it serves as the intellectual superstructure to which major actors in the international order look for legitimation of their interests in ways which are ethically, politically, and socially acceptable; and it is the font for ideas which can help those same actors adapt to pressure for changes. The world intellectual community is thus the privileged locus for floating trial balloons which test the winds of possible opposition movements. Paradoxically, most members of this "invisible college" are highly altruistic and hold highly ethical personal views; they have a deeply felt regard for saving the planet from destructive war, ecological dissolution, demographic catastrophe, mass starvation, and urban decay. Nevertheless, the interplay between the funding requirements of such a system which serves as the legitimating filter of "reputable opinion" and its stylistic congenialities with top corporate, governmental, and international elites does much to harness these noble sentiments to the cause of assuring that change processes remain under the social control of old elites and whatever elements of the "new elites" the old can "live with."

Hence, although it is itself neither homogeneous, fully visible, nor very wealthy, this knowledge coalition has unbridled access to power, riches, and influence.[28] Its intellectual activities play important roles in the functioning of world orders: a maintenance role and an insinuatory guidance role for aspiring stewards of the transition to an altered system. Nevertheless, it remains possible for new loyalties and coalitions to be formed, this time in defense of genuine developmental possibilities.[29] The importance, present and potential, of these actors in the world system is incontestable. Of all those whom Galtung terms *nonterritorial actors*—international governmental organizations (IGOs), international nongovernmental organizations (INGOs), and business international nongovernmental organizations (BINGOs)—members of the world knoweldge industry are the least wedded to the territorial imperatives of the present global system. Consequently, although they interact habitually and congenially with nation-states and corporate powers, their capacity to alter their own organizational structures in response to human needs is considerable. Galtung cites the Pugwash Conferences and the International Peace Research Association as examples of organizations which have properly internationalized scientific interest in war and peace.[30] The important fact, he concludes, is "whether their members are dominated by national values and loyalties or by more universal international loyalties." Also, whether their image of the desired future world assigns major decisional power to a new international elite (of which they will surely be a part) or whether they are committed to the principle of "global populism."[31]

World Communications System

The world communications system is a fifth link in the global chain of developmental networks. Although it is eminently visible—comprising transportation and communications facilities of every type—its operations are not tightly coordinated. The system is a twentieth-century functional equivalent of the imperial Roman road network, a kind of *preparatio evangelica*, not for the missionary diffusion of the Christian religion but for a world order based on reciprocity in exchanges. The "technological unification" of the world has been achieved by modern communications and transportation, which have telescoped, if not fully abolished, time and space. Images of the good life, of social-change strategies, even of possible patterns of the future are rapidly diffused throughout the globe. There exists, indeed, a special political economy vehicled by global image industries, whose main effect is to make world news of local events and convert local events into international happenings.[32] Modern communications are a nonterritorial actor in the global arena; there is always a television camera watching whatever happens in the world. In a second moment, numerous citizens of varied national societies watch in turn. If the long-term result is not necessarily to create in everyone a global consciousness, at least it is to produce in almost everyone a consciousness of the globe. But the impingement of the global communications network is not purely psychological, for it directly affects the content and style of many national activities—education programs, recreational content, the management of news. An illustrative example of this influence is offered by COMSAT (Communications Satellite Corporation). As one evaluative report puts it,

> a single object—the communications satellite—can change the lives of millions of people. Floating thousands of miles above the earth, it can carry telelphone conversations, telegrams, and television programs to places where modern communications are now a distant dream.[33]

INTELSTAT (International Telecommunications Satellite Organization), in which COMSAT (basically a US joint-venture corporation) has a 33.6% equity, views its operations as taking one further step in a process launched over a century ago by the International Telecommunications Union. It removes artificial (that is, national) barriers to communications by allowing for "long-distance exchange of messages without regard to political boundaries."[34] Thanks to its earth-stations in numerous countries, the INTELSTAT/COMSAT system has already influenced the content of national literacy programs, recreational programming, business-exchange systems, and political reporting. At geopolitical levels of the highest dramatic import, the installation of "hot lines" between Moscow and Washington testifies to the important changes wrought in the conduct of politics and diplomacy by a technological facility which is part of the world communications system. It is no exaggeration to state that other actors in the international system would be rendered unable to operate as international actors in the absence of the world communications system as basic infrastructure. With this infrastructure, not only does global power-wielding acquire a new capacity to be diffused but also competing models of human living—whether those of the primitive Tasaday in the Philippines or the communal hippies of the United States—become known throughout the globe. By all known measure of "internationality" (membership, financial participation, arena of cooperation, vision, interests, and program content), the global communications system is an important performer on the stage of international affairs. One hundred and seven countries, territories, or possessions were in 1975 leasing telecommunications services and facilities on a full-time basis from COMSAT.[35] One measure of the unit's success in its first ten years of existence is given by its president in these words:

> During a time when other international relationships have been marked by severe strains, when historic alignments have weakened, and when international economic relationships have been subject to the greatest uncertainty, INTELSTAT has grown and matured, has become more cohesive, and has achieved great stability.[36]

This amounts to saying that global communications will continue as influential actors in the international system.

Values of These Actors

More significant than the global role of extraterritorial actors are the divergences in values and interests which pit them one against the other. Thus one transnational corporation will wage war against another, yet never beyond the point of threatening the existence of the world-market system. Enlightened companies whose leadership is alert to evolving trends in the world order can flexibly make their peace with nationalist or socialist regimes and with governments which would regulate their activities. All companies share an interest in preserving the international market as the central institution of transnational exchange, although some might agree to subordinate market mechanisms to some organizing principle such as world planning for basic needs or a transnational consensus on investment and technological research. On balance, the values and interests of TNCs are congenial to those of the big powers, whose preferred image of the world rests on the pillars of spheres of influence, balance of power, and elite guidance.[37] For the most part, the international scientific and intellectual community has been content to favor better models for designing the future instead of radical change in the current international order. Apart from a few notable exceptions the majority of internationally oriented scientists, although they sometimes compete with governments and international governmental agencies for a voice in decisions, have refrained from challenging the legitimacy of the present international order.[38] One reason, according to Diana Crane, is that

> scientists do not control the financial resources which support their activities. They must continually negotiate for funds with politicians who may be favorably disposed to their cause one day and negatively the next.[39]

World communications interests are powerfully committed to growth and to expanded coverage. From March through December of 1974 the number of earth-station antennas for sending and receiving signals via satellite in commercial service rose from 80 to 107, and "continued growth is forecast."[40] Interests converge around the dream of circumventing, if not abolishing, all "artificial" boundaries. For communications and transportation interests, national boundaries are a two-edged sword. On the one hand, they allow charging usage and installation fees on grounds of national locations. On the other, these interests rely on fissures in these same boundaries to justify their special contribution to the economy of message exchanges.

All global actors perceive that the present international economic, legal, and political orders are undergoing rapid change. Each in its own way is trying to adjust to probable directions of change so as to survive, maintain influence, and protect its interests. Organizations cannot act otherwise, but if the world system is heading towards unknown shores, two questions become crucial: What is the shape of the future order, and who will be the stewards of the transition?

Stewards of the Transition

Several future models vie for the loyalties of critics. Alternative designs of world order run from a unitary world government and homogenized global society unified by modern technology to a pluralistic world order founded on the basic thesis that "the good society is characterized by the co-existence of many social groupings."[41] Countless scholars and institutions now urge that the entire world be reshaped to prepare desired futures. Underlying all these efforts is the assumption that only a systems approach allows one to understand and predict social wholes. We are bombarded with "models" and simulations which incorporate as many as 100,000 variables into their equations.[42] A new vocabulary comes into existence, posing new dichotomies—undifferentiated versus organic growth, logistic versus exponential growth, stable versus disequilibrating growth. Behind these prodigal expenditures of intellectual energy lies the universal quest for wisdom and simplicity in the midst of complexity. The question finally becomes: What are the foundations of human hope? Is the future worth waiting for? Doubts abound because of the ambivalent character of technology itself: Its very promise is uncertain. Lincoln Bloomfield explains why:

> By 1970, technology had produced so many more problems than society could consume that no one doubted that, in Tom Lehrer's lyric words, "If the Bomb doesn't get you, the Pollution will." . . .
>
> Electronic computers for handling information will affect the lives of even the poorest in the world. These means will tighten the world's sense of community. But it is certain that they also will bring danger. . . . There will be worldwide impatience that the living standards of poor people are not better, and an irresistible demand for still faster technological improvement

> aimed at giving all mankind a standard of living approaching that of Western Europe and America.[43]

The future international order lies under the multiple threats of nuclear annihilation, resource disruptions, biological and demographic catastrophe, and ecological disaster. Stewards of the present order—political leaders in the great powers, executives in transnational corporations or international agencies, global intellectuals—generally agree with Henry Kissinger, that there is "scarcity of physical resources and the surplus of despair." Kissinger urged his hearers to forego confrontational approaches between haves and have-nots:

> Whatever our ideological belief or social structure, we are part of a single international economic system on which all of our national economic objectives depend. No nation or bloc of nations can unilaterally determine the shape of the future. If the strong attempt to impose their views, they will do so at the cost of justice and thus provoke upheaval. If the weak resort to pressure, they will do so at the risk of world prosperity and thus provoke despair. The organization of one group of countries as a bloc will sooner or later produce the organization of potential victims into a counter-bloc.[44]

Confrontation is similarly condemned by corporate spokesmen. Economist Neil Jacoby articulates the mainstream corporate view in these words:

> The multinational corporation, able to assemble resources and to organize production on a worldwide scale, has evolved in response to human needs for a global instrument of economic activity. As it evolves further in this direction, it will find itself increasingly frustrated and constrained by national governments. The outcome of this conflict will depend upon the nature of the future world order. . . .
>
> The multinational corporation cannot thrive in a regime of international tension and conflict. The instrumentality of multinational business is man's best hope for achieving political unity on this shrinking planet.[45]

Elite groups benefiting from the existing order fear confrontation and prefer accommodation to new demands on peaceful terms. Weaker groups, however, are wary of pleas for cooperation, viewing them as efforts by the powerful to defuse the pressure for changing basic structures. They agree with Algerian President Boumédiene, that it is necessary for

> Third World countries to create national and international conditions such that the existing relationships of domination could be replaced by just relationships founded on equality and respect for the sovereignty of states, . . . [and] the international community can guarantee the establishment of a new, more just and more balanced economic relations.[46]

Within Third World circles it is widely believed that just relations cannot be established unless weaker partners in the international order ally vigorous solidarity to a combative spirit. The battle is on to shape the future order, and, in the words of Kenyan political scientist Ali Mazrui,

> at the heart of this question is the old issue of equality, which in history has always been linked to the tensions of interdependence.[47]

Although destinies of all societies are linked, controversy centers on the quality of that interdependence. Will it be hierarchical or symmetrical, in the mode of domination by the few or reciprocity among the many? It is unless to invoke the miracles of abundance which could be wrought by technology to assuage the misery of the masses. As Mazrui explains,

> technology, by increasing the inventive and productive capabilities of these societies (i.e., England and America) way beyond those attained by others, initiated a process of massive disparities of income and power among the nations of the world.[48]

Only political action at the international level, buttressed by parallel actions in multiple national arenas, can offset these disparities. This political imperative gives rise to competing models of the future world order.

Richard Falk identifies nine possible new world orders representing both countertendencies in the current system and building blocks for the future.[49] But with an eye toward synthesis, he reduces the roster to three competing models.[50] These are: an approach based on an expanded club of big powers; another based on an appeal to the ideology of transnational corporations which treat the entire world as a market; and a third model called "global populism."

The Great Powers

The first major tendency toward globalist unification rests on a Darwinian or Spencerian notion of the *survival of the fittest*. Unification of world problem-solving, if not of the world itself, should take place under the guidance of the great powers. The club of great powers may have to be expanded to include new aspirants (a nuclear China or the "newly rich" OPEC countries), but world guidance over change processes remains predicated on spheres of influence and informal (or, in one variant, increasingly formalized) consultation among powerful actors. Many Third World leaders fear that this approach legitimates the tacit or overt claims of the rich and powerful to speak *to* and *for* the world, not *with* it. Under this model the poor and weak will be provided for,[51] but they will not share power: the unification of the world will be wrought on the principle of hierarchy rather than equity. As Falk writes, "This is the Nixon-Kissinger-

Brezhnev design for a new world order. . . . The diabolical brilliance of the Nixon-Kissinger foreign policy is to transform nation-statism while preserving its worst moral defects without eliminating its ecological vulnerability."[52] Big powers, in this view, neither can nor ought to abdicate any more of their sovereignty than they have to in order to assure that the evolutionary process does not basically weaken their influence in decision-making circles. Demands issued by less-developed countries are to be labeled *the threat from the Third World.*[53] Even collective requests made by Third World governments in the United Nations are considered irresponsible or dangerous.[54] Industrial nations are willing to share technology and abundance with less-developed nations but will struggle to keep their supremacy in the distribution of political and ideological power. Power and ideological mastery are not to be transferred on the same terms and in the same manner as economic progress or scientific know-how. Yet at work in the world are two change processes which are interrelated: processes concerning production, mastery over nature, rational organization, and technological efficiency, on the one hand, and those relating to structures of power and control over dominant concepts and ideologies on the other. Although both processes were launched by countries now labeled *developed*, they have spread their effects to all societies. If powerful developed countries perceive that their own self-interest requires some sharing of the benefits related to the first category of processes, they will be flexible and accommodating. Under pressure from below they will also make concessions in the domains of power and conceptual legitimacy. Nevertheless, the preferred model of their hierarchs is to guide the transition in ways which give them maximum control over the speed and direction of change. This is why rich-world intellectuals and policy-makers champion the piecemeal, issue-by-issue "problem-solving" approach, instead of one which focuses on overall structures.

Many lesser powers are willing to entrust the transition to a new world order to the great powers. Because they fear nuclear warfare, ecological catastrophe, and disruption in patterns of geopolitical decision-making quite as much as the big powers, they support small adjustments entailing no basic changes. They may on occasion dissent from superpowers on issues such as jurisdiction over seabeds, rules for foreign investment, or the price of raw materials. But their complaints are confined to what Brzezinski calls *instrumental* and not *fundamental* dissent.[55] Their ultimate aim is merely to improve their own bargaining stance within existing parameters. Many opinion-makers in weaker countries see these arrangements as the best they can get. Others become persuaded that the "balance of power" lets them use the "shield" of big powers to protect themselves against their own enemies. Still others view big-power balance as indispensable if nuclear war is to be avoided. Akin is the notion that mass starvation or economic misery in their own societies will not be tolerated by great powers, if only because they wish to prevent the total breakdown of an order which operates to their advantage. There lurks behind such thinking the unstated fear that the burden of an international order in which small nations would share major responsibilities is too heavy for them to bear. Nations are no less prone than individuals to "escape from freedom."[56] Freedom's burdens are indeed heavy; the prospect of participating in a new global enterprise is understandably intimidating to lesser powers.

Another category of less-developed countries endorses a pluripolar model of evolving world order because they themselves aspire to become big powers. This is manifestly the case of India, Brazil, and a few others.

Many intellectuals assume that big powers will continue to play a decisive role in world affairs *if* they adapt imaginatively to new circumstances. But scenarios of big-power control are being challenged by other models for the transition. One such model vests its hopes in the unification powers of the transnational corporation. According to this view, the world will be made one thanks to the depoliticized problem-solving of TNCs whose specialty consists in managing technology worldwide to solve all problems.

Transnational Corporations

"Unification by economic globalism under the auspices of the transnational corporation" is no nostalgic evocation of the British East India Company or the Hudson Bay Company, precursors of today's transnational giants. On the contrary, much sophisticated thinking goes on in corporate circles as to the proper role of markets in global exchange. Hallowed formulas are now repudiated. Jack Behrman states that to continue invoking the cliché "let the market decide" is "to avoid the recognition that it will merely perpetuate the injustices which exist already; it cannot of itself produce justice. But a just system can employ market rules of implementation quite effectively."[57] In the same vein Henry and Mabel Wallich assert that the "ultimate decision about what is to be produced, which is the essence of economic life, cannot be made by the owners, whoever they are, so long as an economy operates in a market system."[58] Their plea aims to convince business managers that neither the survival of capitalism nor the continued profitability of corporations depends on continued private (or "corporate") ownership of the means of production; all that is needed is the maintenance of markets and price mechanisms. The Wallichs claim that growing Third World nationalism, the trend toward socialism, and the nationalization of basic resources will impose their vision even upon traditional corporate thinkers and managers who will have to rally to the new view if they are to survive as profit-making organizations in what, to them, is an increasingly hostile world. Profits will be legitimated not, as in the past, on the basis of economic efficiency but on the ability of TNCs to be socially responsive as flexible "problem-solvers." Corporate spokesmen exercise themselves in "proving" that profit is compatible with justice, ecological integrity, political sovereignty, and other developmental values. Yet it is a fact that the "revolution of rising frustration"[59] is undermining many conventional corporate claims. As a result, corporations are being forced to initiate far-reaching changes in the rules which apply in the making of profit in poor countries. Notwithstanding the assaults upon their legitimacy, however, TNCs continue to argue that "global corporations are the first in history with the organization, technology, money, and ideology to make a credible try at managing the world as an integrated unit."[60] Their claim, in short, is that they alone can master the global organization necessary to administer the planet in beneficial ways. Aurelio Peccei, director of Fiat, does not hesitate to declare that the global corporation is the most powerful agent for the internationalization of human society. Nevertheless, TNCs are alert to the coalition of forces trying to impose restraints on their own creative adaptation to changes in the world system. They also perceive that "ultimately regulation of the multinationals will depend on the development of multinational political institutions."[61] Critics of TNCs judge new political institutions to be necessary because national political instruments are ineffectual against organizations for which national boundaries either do not exist or are seen as "artificial" lines to be circumvented by global planning. Quite logically, TNCs seek an international order which would allow them to operate otherwise than as mere enclaves in poor countries. "Enclave status," to cite A.A. Fatouros, "arises from the superior sophistication in production and management—and, one might add, in planning—enjoyed by such firms over the surrounding economic actors. Quite simply, LDCs do not have the same level of technological and managerial skills.[62] These discrepancies guarantee that transnational corporations will long retain their "foreign" character. If they wish to disarm their critics, they must get the rest of the world to agree to a new definition of what is "foreign." If the entire globe is accepted as the basic unit of human activity, TNCs are no longer "foreign" to anyone. *What TNCs seek in the evolution of the world order is not necessarily hegemony but a new basis for legitimacy.* So long as development and problem-solving are perceived by influential decision-makers as issues to be solved *mainly* by political, ideological, or military means, transnational corporations will judge that their influence is being curtailed for reasons extraneous to the very problems those outdated means are trying to solve. They view the entire world as a vast market. And who can best respond to market signals and market controls if not large global corporations?

Criticism of TNCs which implies that they are at the service of rich-country governments is largely irrelevant. US-based corporations have bribed government officials in Korea, Iran, Bolivia, and Saudi Arabia; they have also subverted internal political processes in Chile and Guatemala. Nevertheless, TNCs do not pledge their primary allegiance to their "home" countries or governments; their loyalties extend to the market as an institution. The more international is the market, the better. Carl A. Gerstacker, chairman of the Dow Chemical Company, dreams of seeing the world headquarters of his firm on some neutral island owned by no nation:

> If we were located on such truly neutral ground we could then really operate in the United States as U.S. citizens, in Japan as Japanese citizens and in Brazil as Brazilians rather than being governed in prime by the laws of the United States.[63]

National governments, in home and host country, are not required to fade away but simply to assume second place. The dream of transnationals is nonterritorial legitimacy in the conduct of pluriterritorial activities. The panic sometimes displayed by TNCs over criticism showered upon them can be traced to their clear understanding that more is at stake than mere economic survival. Just as transnational corporations were getting their first taste of global power, they lost faith in the ability of other institutions to "solve the world's problems." But they think that they can save the world from wars, misery, social chaos, and alienation. If transnational corporations are imperialistic, it is not because they exploit people economically but because their inner dynamics lead them to usurp decision-making legitimacy for global society at large. Paradoxically, their effort is predicated on giving up political power and minimizing traditional political considerations. Corporate visionaries are veritable utopians. Their best spokesmen willingly admit that abuses should be corrected and that market mechanisms must be subordinated to other goals.[64] But they are no less convinced that history singles them out as a new aristocracy, indeed as the only elite able to exercise global power responsibly and effectively. Of course transnational managers will gladly share power with the United Nations, with national governments, and with other international actors—territorial and nonterritorial. But the value system operative in transnational corporate circles implies a quest for a universal mandate to tackle human problems.

In this sense TNCs offer a second paradigm for the evolution of world order away from a state sytem and toward central guidance, under their benign and discreet hegemony. The thirst for a mandate explains the need felt by TNCs to persuade others that profit *need not* be exploitative, that large size need not be necessarily evil, that technological superiority is the very condition for abolishing misery—in short, that TNCs are better equipped than other organizations to preside over the transition to a new world order.

The Global Populists

A third paradigm for the transition waves the banner of "global populism." This term rests on an analogy drawn from national politics: an appeal made by leaders over the heads of power-wielders and directly to the masses affected by decisions.[65] Populism is often a cloak for demagoguery and the manipulation of mass fears and hopes by "charismatic" leaders. But populism can also represent a genuine respect by leaders for the masses and their authentic values. Populism as a political philosophy rejects the notion that leaders should take decisions in the name of people; instead it holds that people should be helped to take their own decisions. Quintessential populist leaders define their own mandates in function of "the people." But who are "the people," and are they able to make decisions on such complex matters as development? The late Paul Hoffman, then administrator of the United Nations Development Programme, believed that

> development cannot and should not be the exclusive province of the "experts" no matter how skillful or well-intentioned. It is too big, too complex, too crucial an undertaking not to merit the involvement—or at least the concerned interest—of the majority of people in every country on earth.[66]

Present transformations in the world order likewise constitute too big, too complex, and too crucial an undertaking to be left solely to experts. Fouad Ajami sees the "top-down" view of the world as undemocratic, unrepresentative, too closely linked with international violence, and too indifferent to universal justice to be acceptable: "Great power policies and visions cannot be said to be in harmony with the interests of less powerful members of the system."[67] But can any working model of a new world order possibly be formulated with a populist orientation? Even if such a model were desirable, must it not get beyond sterile denunciation of present global shortcomings or of elitist alternative models of transition?

Global populism must prove itself able to solve contemporary crises and offer reasonable hope that its instruments of change can work. Unification by global populism entails a commitment, in Falk's words,

> to deal urgently and equitably with problems of war, poverty, environmental decay, depletion of resources, and deprivation of human rights, through the mechanisms of coordination and planning organized around a guidance rather than a governing system. The long-shot possibilities of global populism appear to offer the only alternative to a new wave of neo-Darwinian statism or a planetary takeover by the multinational corporations.[68]

Global populism calls for a coalition of nonelite forces, allied to elites who "defect" from their class values, to struggle against the "massification" of all human decisions.[69]

To counter the evolution of world order under the hegemony of the privileged few, Falk calls for a "normative challenge" to serve as counterforce to the pursuit by the powerful of their vested interests, all in the name of realism.[70] There is no need to review what a strategy for transition in the mode of global populism requires.[71] But it is important to note that elitist stewardship of transition is neither inevitable nor salutary. Various Third World actors must be mobilized to resist trends favoring great powers or rich institutions. So must intellectuals and professionals—including the military—in rich countries; counterculture and dissident movements in many lands; and institutions such as churches, foundations, and think tanks. Success demands that radical political actors support, and in turn gain support from, innovative social scientists, systems engineers, and futuristic "problem-solvers" of all types. There is no single "focus" or high ground upon which all anti-elitist efforts can be situated.[72] The most fertile soil of such efforts is the mass of the populace. But in societies characterized by multiple overlays of institutional intermediaries between "the people" and leaders, linkages must be established at every level if "the people" are to gain effective leverage over the decisive institutions at work in their societies.

The "third option" for a new world order—global populism—faces enormous odds. In order to succeed it must overcome, in reformists, a two-fold impotence: the defeatist illusion that they can do nothing in the face of overwhelming trends and the romantic utopianism of those who paint static portraits of desirable alternative worlds.[73] The great powers, supported by timorous lesser powers, may, it is true, be able to exercise stewardship over the transition to a new world order in accord with the twin values of elite control and stability management. And chances seem good that transnational corporations will prove able to parlay their economic and organizational power into political gains. One readily imagines circumstances in which traditional political instruments will so dramatically reveal their helplessness to manage global problems that TNCs might easily assume *de facto* hegemony through sheer default. Were they to do so, however, they would surely delegate many specifically *political* tasks of governance to those same powerless institutions they had just supplanted. This phenomenon often occurs when military regimes seize power from "incompetent and corrupt civilian politicians." Military elites quickly discover that they cannot govern alone; they are forced to woo technocrats and politicians back into action. It is neither the politicians nor the electorate which calls the tune, however, but the military. Overlapping interests evidently link the destinies of large governments to those of transnational corporations, and mutual accommodation rarely proves impossible. Nevertheless, the dynamics of global profit-seeking create pulls in opposite directions than those generated by the pursuit of global power politics. To illustrate, the US government fears that ITT's private war against Salvador Allende in Chile eroded its own ability to conduct foreign policy. More ominously, the American government begins to fear that it cannot control arms sales to potential enemies by corporations within its jurisdiction. A recent editorial states the danger in these terms:

> There must be a way to ensure that the U.S. government applies substantial political criteria to arms sales to oil-rich countries and that it does not give its corporations reason to believe that anything they do to their own profit is perfectly acceptable to official Washington.[74]

These fears extend to the domain of nuclear policy. The announcement made by the Bonn government in June 1975 of its intention to supply Brazil with a complete nuclear industry and technology has alarmed Washington, which is already fearful that Taiwan, Argentina, Chile, Pakistan, Israel, and South Africa will soon join the nuclear club. One editorial on "nuclear madness" tells us that

> should Bonn perpetrate this nightmare upon the world,...it will pay a political price that will far outweigh political gains. A much wiser course would be to join the United States in refraining from such sales and in urging other supplier nations to move quickly toward common export rules, rather than the *competitive degradation of safeguards in pursuit of profit.*[75]

Both the pursuit of profit and the quest for geopolitical influence, however, threaten human survival, ecological integrity, and the possibilities of humane development for all. These values, which are not served by the extant world order, are also incompatible with the first two "models" of transition. This is why global populists actively fight an uphill battle in favor of an alternative world order. One cannot be human, they argue, without creating new possibilities. And as Camus wrote, it *is* worth making the supreme sacrifice for the sake of the possible. Therefore, "true generosity towards the future consists in giving one's all to the present."[76]

The stakes in the battle for stewardship over the transition to a new world order are defined by Falk as follows:

> (1) The state system is being superseded by a series of interlocking social, economic, political, technological, and ecological tendencies which are likely to eventuate in some form of dysutopia or negative utopia, that is, in a very undesirable and dangerous structure of response to the problems posed by the deepening crisis in the state system.
>
> (2) Although this disquieting outcome seems probable as of now, it is not inevitable. There is also a beneficial option, premised upon an affirmation of the wholeness of the planet and the solidarity of the human species, that could bring about a rearrange-

ment of power, wealth, and authority that would be more beneficial than anything the world has heretofore known.

(3) Initially, the global reform movement needed to underlie such a positive outcome has to take principal shape outside of and mainly in opposition to the centers of constituted political and economic power—it will almost certainly have to be populist and antigovernmental in character and origins. Such a movement should be premised upon nonviolence to the extent possible.[77]

(4) The principal initial focus of a movement for positive global reform should involve education-for-action, that is, demonstrating that the felt needs and frustrations of people in a variety of concrete social circumstances around the world arise from the inability of governmental or multinational actors to find short-range, middle-level, and long-range solutions to the distresses and dangers of our world.

(5) The case for global reform should be premised on a basic assessment of structural trends and options. It need not rest altogether on the collision course conveniently being programmed by apocalyptic reformers to take effect by the year 2000. We should be somewhat suspicious about the recent show of millennial egoism—either change by the year 2000 or everything is lost.[78]

Because any transition toward a world order congenial to development requires the institutionalization of new allegiances, one must inquire into the dynamics of loyalty systems.

New International Loyalty Systems

A decade ago Edward Banfield evoked the difficulty of transferring the ethical loyalties of people from a narrow family to some larger society.[79] He correctly situated allegiance systems at the heart of ethics. Primary loyalties usually win out when they conflict with other allegiances, even those which are formalized in social norms or public rhetoric. But loyalties are no exclusive attribute of the Sicilian villagers described by Banfield or of "traditional" societies. Parallel tensions in all systems explain the behavior, and probable future responses, even of "modern" actors in the international order now undergoing change. A closer look at the loyalty systems of aspiring "stewards of the transition" helps explain the vested interests they have in certain models of the future order and the probable concessions they will make to Third World technology policies.

The postulate bears repeating: The existence of multiple loyalty systems is *not* confined to "underdeveloped" groups. On the contrary, modern societies have fractured not only the unity of cognitive activities but also the emotional bonds that bind individuals to causes and to other people. Thus specialization of multiple allegiances characterizes "moderns"; along with loyalty to peer groups, they retain residual allegiances to clan, family, ethnic group, or some freely embraced "movement." To the extent that their work makes of them international actors, other loyalties tend to take second place. Yet even in international arenas, many professionals feel obliged to render ritualisitic homage to their national loyalties. As Perroux notes,

> mandatories of states cannot, in any conference or meeting among states, speak otherwise than as partisans and representatives of their states. Properly national leaders deprive themselves of all immediate influence if they consent to speak in the name of an experience or an ideal higher than that of a single nation.[80]

What Perroux laments in mandatories of states holds likewise, in great measure, for nongovernmental actors. Most international meetings are so structured as to make participants unduly conscious of their nationality, their professional discipline, or the ideological system they are deemed to represent. Thus are they made accomplices of loyalties which may not be deeply felt. Deep within, however, professionals concerned with the future world order give their loyalty to a new humanity in gestation and to a future social compact only now beginning to delineate itself. "World consciousness" may already have advanced enough to invalidate the judgment pronounced by Perroux in 1958.[81] It is no longer "internationalists" who are heretics in such arenas but rather the partisans of "narrow" parochial interests. Nonetheless, the abiding ambivalence of loyalties helps explain why the globalism embraced by a corporate manager differs from that espoused by a World Bank official or a Third World intellectual. The key lies in detecting, beneath the formalistic (albeit genuine) national and international allegiances, the *intimate* loyalties.

Interpenetrating loyalty systems are illustrated in the attitudes of self-styled "internationally minded" corporate executives. Conventional wisdom has it that the primary allegiance of such executives is granted to the corporation which employs them or, at the very least, to the corporate system. But this is not an adequate answer. Like intellectuals, bureaucrats, managers, financiers, reporters, and other professionals, corporate officials are more loyal to their profession than to a particular employer. Their unflinching defense of job mobility points to the source of their security: the marketability of their skills, which rests, ultimately, on the judgment of their peers. But, one may object, is not the survival of the corporate system vital to the welfare of international managers? Not really, if we are to believe the private testimony of many of them. In countless interviews, executives have professed their "realism" in adjusting to a changing world. Most of them would not have hesitated to work *for*, and not merely *with*, socialist governments, state enterprises, or other noncorporate employers. What is essential to their professional and personal identities, however, is guaranteed peer-level remuneration, status, and mobility.[82] They will not contract their services to socialist regimes if they are to be paid at egalitarian or "nonprofessional" salary levels established on "ideological" grounds. Moreover, managers of state-owned enterprises often lack prestige in the eyes of the "sophisticated and modern" international business community. And furthermore, successful managers do not wish to mortgage their future in ways which curtail their lateral and vertical mobility; they must remain free to work for others should "irresistible" opportunities arise. Truly international persons of this type are equally at home in Nairobi, San Francisco, São Paulo, or Singapore. Not, of course, that national origins, preferred places of residence for their families, or esthetic and climatic tastes do not create, in them as in others, ties that bind. But their most basic loyalties are to the networks that best assure them the three advantages just named: high salaries, the status of professional peers, and the maintenance of mobility. When loyalties conflict, salary, status, and mobility carry the day over lesser considerations of national policy, ideological fidelity, or gratitude to the company. This very loosening of the bonds of allegiance to nation, family, religion, and locality is what makes transnational personnel targets of suspicion for those in whom nation, ethnic group, or locality still elicits strong fealties. The hero in John P. Marquand's novel *Sincerely, Willis Wayde* at one point comments that "in business, loyalty acquires a new definition every day."[83] This portrait captures a normative tendency powerfully at work in transnational corporate enterprises which socialize "successful" adepts into subordinating local, cultural, and national loyalties to the "larger job of seizing opportunity and making it profitable." Each firm tries to instill loyalty to itself among employees, but the logic of its own reward system is too strong. This system conveys in irrefutable terms the notion that competitive flexibility is the key to success.[84] If the firm as a whole insists on being both competitive and flexible—geographically, culturally, and operationally—then why should not its officials display the same attributes?

National loyalties are on the wane in many arenas outside those of business, thus making it easy for corporate leaders to champion the internationalization of life. In simpler times, many TNCs were content to rely on the political influence of their home countries to protect them overseas. Fissures now developing between the interests of such firms and the governments of their home countries incline managers to envisage a new status for their corporations and for themselves. The network of their primary loyalties obviously runs counter to the dominant interests of other actors in the world arena, especially those of governments.

The durability of big-power influence in decision-making is predicated on the survival of nation-states as primary objects of collective loyalty. Patriotism provides a heady dose of "emotional integration" to populaces the members of which otherwise have little in common one with the other. If too many people "defect" from national loyalties, however, the governmental managers of big powers will lose their base. The ground on which politicians in weak countries stand is no less tenuous: their influence at home and abroad is founded on the maintenance of national allegiances among their compatriots. It thus comes about quite naturally that most governmental elite groups develop loyalty systems which formally relate them to concrete patriotic institutions, all the while simultaneously giving their deepest allegiance to the maintenance of their own power bases. In a world becoming ever more transnational in its images, travel habits, and demonstration effects, the instrumental loyalties given to power bases replace the more rigid local or class loyalties operative in earlier times. The analogue to class consciousness for many people nowadays is the sense of belonging to and receiving the respect of peer communities. Because political leaders can gain that respect only if they also enjoy credibility among masses of poor constituents, they are tied to such values as patriotism, nationalism, ethnic culture, and particular goals. International-agency personnel, on the contrary, are less dependent on these sources for their professional identity, institutional legitimacy, security, or role definition. Their access to constituencies is remote, indirect, and, in the final instance, quite irrelevant to the performance of their tasks. Nevertheless, because most of the institutions with which they negotiate are nation-states, such personnel more readily favor transnational cooperation rather than the dilution of national loyalties favored by corporate executives. This is not to say that individuals in world organizations unconditionally seek to preserve nation-state sovereignty or are hostile to world unification under the managerial hegemony of transnational corporations. But it is to say that the loyalty systems at work in their institutions favor certain models of the evolving world order over others. Their institutional loyalties incline them to a relatively egalitarian consensus or compromise among nation-states, rather than to a big-power-hegemony scheme of global direction, a transnational-corporate-leadership model, or a "global populist" paradigm. Because their own preferences are unabashedly elitist, the "populist" model of world order appears to them both utopian and threatening.

The "global populism" model *is*, of course, utopian: it breaks sharply with known precedents. One major crisis of the present world order is traceable to a growing distaste for elite rule of any type.

In developed as in less-developed societies a sense is growing that no policy *for* the people can be formulated except *by* the people or at least in association *with* them. This aspiration is frustrated by those very systematic characteristics which typify "modernity": the vast scale of operations, their technical complexity, the minute division of labor which results therefrom, the overlapping interdependencies which link local and regional events to worldwide happenings, and the ever-shortening time lag between the impingement of change on societies and the response they must make to assure survival and integrity. At the very time when it has become more difficult than ever before to diffuse decision-making, demands for participation have escalated. Those who voice the demands, however, remain largely impotent to define the instrumentalities needed to satisfy them. Here in fact lies the greatest debility attaching to the "global populism" approach to transition: While in moral terms it is the most desirable, it is also the most difficult to implement. It is difficult even to visualize. Transformation under the aegis of big powers or of transnational corporations, on the other hand, is at least easy to imagine. And although the vision of world order preferred by big powers or by TNCs is not easy to implement, the instruments needed to succeed are already known, and the march of events is shaping the contours of a new order in their favor. Therefore, developing a strategy for creating a world order of peace, equity, participation, and ecological health is clearly an uphill journey against very heavy odds. Notwithstanding its difficulty, such an effort illustrates how local and international loyalties might coexist.

Paradoxically, few people can be enthusiastic global populists unless they also have strong local loyalties; in order to counter the abstract universalism of systemic decision-makers, one needs experiential and existential roots. And one must be kept accountable to a living community of human need, not merely to some model, plan, discipline, profession, or utopian vision. This kind of accountability helps place "experts" in horizontal relationship with others, thereby facilitating their necessary apprenticeship in exercising their specialties in a horizontal mode. Allegiance to a concrete community is no sufficient guarantee, however, of sound globalism. Local or special attachments should never exhaust the objects of one's loyalty. All human beings are members of the same race, the same ecosystem, the same network of living beings. Given the characteristics just named—large scale and multiple divisions of labor—the "micro" realities of our existence can only be healthy if they contribute to a healthy "macro" system. The Bengali poet Tagore thought that, in the present era, only those values which could prove themselves universally valid could be considered genuinely human.[85] Similarly, a Brazilian university rector insisted ten years ago that universal development values can be achieved only by deep commitment to local and regional problems. He accordingly titled his book *The Universal Through the Regional: Definition of a University Policy*.[86] Ways must be found of reinforcing local attachments so as to free people to embrace wider loyalties as well. The obvious danger is that the nurturing of broader allegiances will undermine people's loyalty to their more immediate concrete realities, for these, by definition, are always limited in space, time, and culture. Karl Mannheim, ever conscious of the dialectical tension between the universal and the particular, contrasted alienation, which is pathological and destructive of creative energies, to a special kind of uprootedness which liberates one to be genuine, simultaneously, at "micro" and "macro" levels. Writing twenty-five years ago, he used these terms:

> Neither the place or country of birth, nor the nation in which they happen to live means much to them. We usually call this process uprooting, and the pejorative sense of the term is justified insofar as with most people loss of identification with a definite locale and non-participation in community life leads to disintegration of character. This detachment from a locale of one's own leaves a feeling of belonging somewhere either undeveloped or unfulfilled. It makes for mental insecurity and unattached emotional states, leaving people easy prey to propaganda. . . . [But] what we pejoratively call "uprooting" has its positive aspects both for personality formation and the construction of a world-community. Hardly anybody will doubt that the establishment of larger communities—possibly a world-wide community—is possible only if people overcome the state of unconditional subservience to the power demon of national sovereignty and aggressive nationalism. Partial uprooting, emancipation, is therefore necessary and is indeed achieved by progressive man.[87]

The *mode* of giving one's global loyalties is crucial. If one repudiates local accountabilities or takes elite peers as one's primary "significant others," uprooting from lesser loyalties will prove damaging: it will transmute one's leadership roles into postures of rulership.[88] Moreover, it can confirm the illusion held by many planners and decision-makers at the top, namely, that they know better than the people themselves what is good for them. Some expert planners now repudiate this view on grounds of pure efficiency.[89] And because of growing linkages between domestic and foreign policy in most countries, even local citizens mainly interested in their own needs understand the importance of having a voice in their country's international policies.

Argentine political philosopher Marcos Kaplan sees a need to create unprecedented horizontal power arrangements. Less-developed countries, he argues, face three key problems, and power lies at the heart of each.[90] First, these countries must devise internal development strategies to replace those which have proved unsuitable. Second, they must form new types of relations among Third World countries as a whole. And third, they need to restructure relations between the Third World and the rich world. The major problem is this: Throughout recorded political history, large-scale power has always been vertical and hierarchical, never horizontal. Although one modern *ideology*, namely, socialism, seems to favor new modes of horizontal power distribution, socialism *in its historical incarnations* has usually presented itself as a shortcut for achieving what capitalism already possessed by other means—that is, a strong industrial base, material abundance for the masses, technological modernity, and a generalized sense of welfare. Especially in poor rural societies where political participation had never existed, socialism promised to reach these goals faster and at lower levels of human sacrifice than capitalism. Socialism thus engaged itself in the competitive race for efficiency in "delivering the goods." Inevitably, the way was cleared for self-proclaimed "enlightened minorities" who arrogated to themselves the right to speak for the masses and to organize societal tasks allegedly in their interests. This amounted to the vanguard minority's establishing a tutelage over the masses. Whenever this happened, fatal mechanisms made their appearance in the political arena. What occurred was not, as in capitalism, the economic expropriation of the plus-value of labor input but what Kaplan calls the "political expropriation, by the minority, of political power which, according to socialist doctrine, ought to reside in the masses." By a quasifatalistic process this political expropriation has led to the parallel expropriation of economic resources by those "new classes" Djilas described: party *apparatchiks*, bureaucrats, technocrats, and "useful" professionals.

"Liberal democracies," however, have not been any more successful in providing horizontal access to power, notwithstanding their rhetorical promises. Under capitalism, the market imagery of which powerfully reinforces liberalism's preference for the "free play of ideas" and pluralistic competition in the political arena, economic expropriation took place first and redounded to the benefit of the more successful economic competitors. A representative political system was devised which favored the interests of those at the top of the economic pinnacle. Clerks and bureaucrats were recruited to play roles supportive of these interests. Thus the horizontality of power was denied in fact by the primacy of money.

For Kaplan the major task facing Third World societies and the transnational order itself is to create modes of "horizontalizing" both economic and political power. Success presupposes a high degree of self-management and participation by the public in vital decisions. Therefore, the "global populist" approach to reshaping the world order is not only preferable; it is indispensable if the three crucial problems just evoked are to be solved. The same dynamics underlying the populist path to a new world must likewise be made operative within domestic societies: A loyalty system to the human race at large must coexist with allegiances to local communities which will engage one's energies in the struggle to create effective self-management and diffused participation.

This twin loyalty is incompatible with a purely abstract form of internationalism in which the only loyalties are to peer reference groups (as the ultimate font of status, mobility, and legitimation for high levels of remuneration) and to world markets to be exploited or to power systems to be managed. Populist loyalties bind one simultaneously to larger demands of diverse human solidarity and to those of local communities. The first allegiance protects one against parochial ethnocentrism; the second, against escapist and alienating "uprooting."

Particularly in the case of expatriate bureaucrats and professionals, institutional loyalties have replaced national or cultural allegiances. But this new loyalty system creates two problems: It is elitist, and it is functionally universal with no roots in concrete struggles and a specific cultural identity. Although examples of healthy local loyalties which are simultaneously nonethnocentric and universal, embracing the whole human race, are few, these twin loyalties are indispensable. Countless individuals and institutions must adopt them if an international economic, legal, and political order congenial to genuine development is ever to become possible. (Again, the phrase *congenial to genuine development* implies compatibility with a technology policy which is value-enhancing instead of value-destroying. The reason, as argued earlier, is the vital nexus linking development value options and strategies to such policies.)

A note of explanation is needed here: All institutional actors described herein harbor within their confines some individuals with the requisite dual loyalties. Therefore, institutional actors can change under the impact of human wills. International bureaucrats can discover numerous ways to become accountable to concrete communities of need, even if they have at times divorced themselves from local allegiances to cultural origins. Similarly, enlightened personnel in transnational corporations can, in virtue of their loyalties outside the firm, pressure their own institutions to adopt radically different roles in a future order. In truth, one occasionally meets executives who privately admit that the survival of transnational corporations is not essential to civilization in the future, so long as the corporation's special contributions—managerial skill, flexibility, coordinating abilities, and technological dynamism—are not lost to humankind in its organizational problem-solving efforts. Without going this far, others nonetheless concede that corporations are public institutions the total behavior of which should be subject to public control.[91] In a 1975 speech W. Michael Blumenthal, then president of the Bendix Corporation, and now US Secretary of the Treasury, appealed to the business community to accept dialogue with critics and constituents. "An entirely new approach is needed," he said, "a frankly moral approach that would begin with business taking a long, hard look at itself."[92] The ideas of responsibility and ethics in business in the broadest sense should be translated into practice by the concerted efforts of the business community, lawyers, the clergy, statesmen, philosophers, "and others whose views would represent the moral concerns of society as a whole."[93]

These words are doubtless designed to allow business, in Blumenthal's phrase, "to fend off punitive, heavy-handed and possibly damaging legislation that the public will insist on if a degree of self-policing is not seen to be effective.[94] A second putative benefit is that confidence can be restored to business "by improving the performance, and not merely the image, of our business organizations."[95] This is clearly one form of survival strategy. But in the present crisis even survival strategies need to be ethically sound; otherwise they are doomed to failure.

The pressure of events will force even the most ethically callous institutions to "internalize" many values which they had blithely "externalized" in the past. The words penned by the late Adolf Berle in 1954 take on prophetic meaning nowadays:

> The really great corporation managements...must consciously take account of philosophical considerations. They must consider the kind of community in which they have faith, and which they will serve, and which they intend to help to construct and to maintain. In a word, they must consider at least in its more elementary phases the ancient problem of the "good life," and how their operations in the community can be adapted to affording or fostering it.[96]

Corporations, it is evident, can acquire "bad conscience" only under the pressure of socially organized complainants.[97] The same holds true for other organizations: governments, international agencies, and the world scientific community. They will adopt the global values and behavior championed by "global populists" only if the latter's appeal satisfies the deep need felt by these institutions to assure survival or functional relevance. This exigency, therefore, dictates a strategy patterned after that adopted in other circumstances by revolutionary guerrillas. Although guerrillas understand that they must simultaneously take initiatives themselves and mobilize masses, they also recognize the importance of winning "defectors" from the police, the armed forces, the public bureaucracy, the intellectual institutions, and other fonts of support for the very system they are challenging. The lesson for global populists is clear: part of their strategy relies on appealing to the multiple loyalties of those presently serving global elitist constituencies. This is the locus where loyalty systems become directly germane to the struggle presently going on among aspirants competing to become stewards of the transition to a new world order.

Were hegemony over the evolving global order to fall unchallenged into the hands either of an expanded club of big powers or of a directorate of market-oriented global corporations, the prospects of achieving "humane" technology policies would have suffered an irreversible setback. Technology is tightly correlated to power and patterns of decision-making. If, therefore, world exchange systems become hegemonic or (in Kaplan's terms) "vertical," this will *ipso facto* represent a victory for "big" technology, for "centralized" and expansionistic technology—in short, for precisely those forms of technology which impede genuine development. For this reason the evolving world order is something far more than a mere backdrop against which is fought out the battle of technology: it is rather a central battleground in the struggle to harness technology to human causes. Failure to so harness technology will lead to what W.T. Stace calls "the cosmic darkness," the final victory of the scientific revolution, for "belief in the ultimate irrationality of everything is the quintessence of what is called the modern mind."[98] Irrationality is the absence not of causes but of purpose. And if technology displaces all purpose to assure its own survival, the highest rationality becomes supreme irrationality: only means can exist; ends have become impossible. The universe will have ceased to have any meaning or any *raison d'être* except this one: it provides the experiential raw materials to be managed by technology. Technology is indispensable because the world cannot be managed without it. Technique indeed is well on its way to becoming the only permissible metaphysics,[99] the sole normative value, the universal substitute for culture.

It is important to clarify the relationship between loyalty systems and the ideology of the marketplace. The central tenet of market ideology states that ideas, skills, goals, and values—quite as much as goods, services, and material inputs for production—are best allocated via the machinery of competitive circulation within circumscribed arenas. Two of the three models for a future world order described in the foregoing pages—the big-power and the transnational-hegemony models—have their roots in this ideology. Political liberalism simply assumes that political decision-making should be vested primarily in the strongest, that is, the most *successful*, competitors in the power struggle. Quite logically, therefore, liberalism defines the participation of "the people" in ornamental or incidental terms: The people merely ratify decisions made by rulers or elite advisers and are given sporadic opportunities to replace one set of rulers or advisers by another.[100] The market ideology is endorsed even more openly and explicitly by transnational corporations which invoke their superior efficiency in certain limited economic domains as justifying a mandate to bring about global unification. In terms of the first model, transnational corporate activity is necessary to success but is subordinated to geopolitical considerations. Under the second hypothesis, hierarchical roles are reversed: world politics need to be "disciplined'' by the exigencies of economic efficiency as it is practiced (if not as it is defined) by transnational firms. Only the third pattern of a possible future world order, that designated here as *global populism*, repudiates the market ideology. Instead of assigning decisional power on the basis of competitive success, it grants priority to human needs; to the inherent dignity of all individuals and cultures; and to higher values of ecological health, human survival, and optimum justice. Conflicting perceptions as to the legitimacy of the market need to be clarified if one is to grasp the relevance of competing world-order models to technology policy. This is doubly true because much criticism of current technology exchanges points directly to abuses or defects in market mechanisms.

A Note on Markets

One economist laments that ''under present turbulent and rapidly changing conditions, our ability to focus on ends is seriously jeopardized.''[101] He further declares that one must identify desirable ends in clear terms and recognize that certitude as to whether these ends are ultimately achievable is secondary. He then concludes that ''the measure of success might be found in the efforts exerted toward achieving the goal rather than in reaching the goal itself. The end, in the final analysis, therefore, might be the effort itself of continual examination of processes by which ends are sought after.''[102] This same analyst sees the immediate problem of the nations of the world as a resource question. Can the market mechanism, he wonders, determine resource allocation and distribution, or will a gigantic power play ensue in which the survival of the fittest will prevail over the human quest for justice, equity, and sufficiency for all? One need not look far for an answer. Eugene Skolnikoff declares flatly that ''it is beyond question that present market mechanisms cannot adequately represent all the objectives of a society.''[103] The crucial problem is that markets have always been defended by their supporters as being efficient tools for allocating resources, distributing income, and organizing the tasks of consumption and production. So long as discussion was thus centered on issues of ''efficiency,'' economists and managers could exclude values from the efficiency equation. New global tensions, however, and higher consciousness of mass misery now force us to ''internalize'' such former ''externalities'' as the abolition of need, ecological integrity, and the elimination of alienation in production lines. The vital question needs to be rephrased: Is the market still efficient *if* it must internalize such values in its cost-benefit calculus?[104] No answer can ignore the important distinction invoked by Karl Mannheim regarding two different roles played by the market. He writes:

> Competition and co-operation may be viewed in two different ways: as simple social mechanisms or as organizing principles of a social structure. . . .
>
> This distinction may help to clarify the question whether capitalist competition—allegedly basic to our social structure—need be maintained as a presumably indispensable motivating force. Now, one may well eliminate competition as the organiz- *principle* of the social structure and replace it by planning without eliminating competition as a social mechanism to serve desirable ends.[105]

What Mannheim says of competition can be applied to the arena within which competition unfolds, the market. The market may be viewed as the organizing principle of economic organization or as a regulatory mechanism. Even under neocapitalism or welfare capitalism, nonmarket mechanisms (such as welfare legislation or corrective taxation) can do little more than correct the worst abuses of the market system.[106] But this is no justification for allowing the pendulum to swing to the opposite end—abolishing markets. Local, regional, national, and global markets must be preserved. But they must be subordinated to some new *organizing principle* for allocating resources and setting the objectives and modes of production. This organizing principle needs to be founded on a new global compact or social contract around priority values like survival, justice, equity, sufficiency for all, ecological integrity, and the elimination of large-scale systematic violence from human life. Left to its own inner logic the market cannot assure, or even allow, the attainment of these goals. The logic obeyed by markets and competition (as organizing principles of economic activity) provides rewards in directions opposed to the humane values just mentioned. That logic feeds armaments races, ecological ravage where great profits are to be had, inequities which respond to the effective purchasing power of the rich, and the race toward biospheric death in obedience to the "technological imperative" described earlier. Some contend that even profit-makers can be socially responsible and sensitive to human values.[107] They add that politicians likewise seek peace, survival, and justice and aspire to be "statesmen," architects of human progress, and not mere manipulators of limited interests. These claims are partly true, but existing interconnections among all major problems and societies guarantee that these desirable goals cannot be reached unless the pursuit of power and profit are subordinated to higher values. Market mechanisms and political competition must be dethroned as organizing principles and transmuted into regulatory mechanisms at the service of a global ethic of need and value. These mechanisms will no doubt retain important functions: to control against waste, excessive centralization, arbitrary imposition by experts of production targets, duplication of effort, and inefficiency. But ultimately a new global order must assure access to resources for all individuals and societies on the basis of need and *independently of political strength, of market competitiveness, or even geographical location of the resources.* Access to resources needed to provide material sufficiency at a modest level must be guaranteed *upstream.* This means that it is not acceptable simply to have finished goods redistributed by those few who effectively control production. The very choice of how to process and use resources must be widely shared. Galbraith once wrote that "the final requirement of modern development planning is that it have a theory of consumption... a view of what the production is ultimately for... *more important, what kind of consumption should be planned?*"[108] All human groups need an opportunity to decide which priority consumptions they should plan for. This is why it is not enough for them to have access to resources *downstream* after initial decisions have been made by other producers.

The principle that resources for elementary needs belong to the whole race collectively will be attacked by many Third World and rich-world observers alike. The former so insistently assert their sovereignty over resources found within their borders that the proposal will seem regressive to them. Nevertheless, the poorest Third World countries are discriminated against by erecting purely geographical sovereignty over resources into an absolute principle. Rich countries, in turn, will view the principle of *upstream* access on grounds of need and equity as a direct threat to their present affluence and competitiveness. But in the long term it is sheer palliative to imagine that economic justice can be founded on any other principle but one assuring such access. Either a global ethical agreement about the relationship between all resources and all truly basic needs is brought into existence or other systems will merely correct the worst evils of the principle they adopt.

An irreducible tension exists between concern for public developmental achievement and private business. Corporate consultant Thomas Aitken points to this tension when he asserts that "the overseas manager often finds that he is concerned about public affairs only as they affect his business; otherwise he has become uncommitted."[109] The historical record of development planning is strewn with cases wherein political priorities led leaders to compromise goals of improvement for the masses in order to stay in power or crush competitors. Consequently, it is unrealistic to deposit any hopes for a world order congenial to development and to humane strategies of technological use except in something resembling the "global populism" approach outlined by Falk and his colleagues. As he himself acknowledges, if the approach should fail to achieve total success, it can at least mitigate the damage likely to ensue if either of the first two competitive models is adopted. Issues of priority concern to the Third World must be attended seriously by all claimants to the role of "stewards of the transition." As Soedjatmoko, the Indonesian philosopher and diplomat, writes, "the moral legitimacy and persuasive power of any concept that may be formulated by North or South will depend in large part on where the poor and resource-poor part of the Third World, the so-called Fourth World, with both its problem of poverty and its potentialities, fits into the scheme of things."[110]

The same must be said of technology and the multiple problems it poses: these cannot be solved unless the resource-poor world "fits into the scheme of things." The ability of the world at large to harness technology to the needs of the poorest may well prove a decisive touchstone of survival and social evolution in the human species. Value conflicts encountered in technology transfer are but a pale reflection of deeper conflicts lying at the heart of every society's effort to make technology serve human goals. Technology readily elicits a spectre of a robot or a Frankenstein monster: the creature which overwhelms its human creator and destructively masters its progenitor. The future of humanism itself depends on humankind's ability to tame that beast, to control that mechanized "animal." André Malraux's warning seems well-taken here:

> Humanism does not consist in saying: "What I have done no animal could have done," but rather: "We have refused to carry out what the beast within us would have us do, and we are determined to rediscover the human at all those places where we find what crushes the human."[111]

Conclusion

Conclusion

Neither past cultural traditions nor the present scientific mentality, *taken separately*, can supply a "wisdom" for harnessing technology to humane ends. Margaret Mead sees the future as "the appropriate setting for our shared worldwide culture, for the future is least compromised by partial and discrepant views."[1]

In truth, however, the future *is* compromised by "partial and discrepant views": competitors vie for control over the evolutionary process toward a new future. They place different values at the heart of culture;[2] yet none can avoid asking two perplexing questions:

(1) Can technology be controlled?
(2) Is technology compatible with civilization?

Controlling Technology

The most radical analyst of technological determinisms remains Jacques Ellul who, ten years ago, stated his belief that

> the technical phenomenon has assumed an independent character quite apart from economic considerations, and that it develops according to its own intrinsic laws...from man's intentions, following its own intrinsic causal processes, independent of external forces or human aims.[3]

Although accused of determinism, Ellul retorts that he "never intended to describe any inexorable process or inevitable doom." On the contrary, he insists that

> if we can be sufficiently awakened to the real gravity of the situation, man has within himself the necessary resources to discover, by some means unforeseeable at present, the path to a new freedom.[4]

More recently Ellul has reaffirmed the possibility of controlling technology in terms of

> conflict between hope and the dominance of technology. The latter can neither tolerate the future-eternity relation nor the inter-

> vention of a future composed in the present. Technology is expressed by means of necessity through cause and succession. It is incapable of entertaining any other prospect. We now are called to another prospect, but that in no way implies a condemnation of technology! It implies simply the observation that salvation is not to be had from that source, since technology unstructures time and blocks the movement of hope.
>
> What we have eventually to do as Christians is certainly not to reject technology, but rather, in this technological society and at the price of whatever controversy, we have to cause hope to be born again, and to redeem the time in relation to the times.[5]

Ellul's point is that technology cannot be controlled if one assumes it is easy to control. Bertrand de Jouvenel condemns this view as "impotent and paradoxical technophobia." A pioneer in the study of "futuribles" (future possibles), he declares that

> membership in a technologically advanced and advancing society is unquestionably a privilege. It is true of all privileges that they can be put to good or bad use. In this case it is quite clear that the privilege is collective by nature, that is the benefits and the evils depend a great deal more upon aggregate behavior than upon individual decisions.[6]

The monumentally important fact is that the whole world seeks membership in this society, although privilege is not automatically conferred upon new national entrants to the technological club. Aggregate behavior will no doubt decide whether technology will be harnessed to human ends or be allowed to subvert those ends. *Subversion* is not too strong a term, for as Everett Reimer notes, "science and technology violate nature, including human nature."[7] But technology *need not* violate nature: it will simply continue to do so unless humans force it to impinge differently upon nature. To achieve control over determinism, humans must first free themselves from their hypnotic fascination with technology's benefits. As Ellul writes,

> all men must be shown that Technique is nothing more than a complex of material objects, procedures, and combinations, which have as their sole result a modicum of comfort, hygiene, and ease. . . . Men must be convinced that technical progress is not humanity's supreme adventure. . . . As long as man worships Technique, there is as good as no chance that he will ever succeed in mastering it.[8]

A few prophets, philosophers, and poets have remained, it is true, immune to the idolatry of technique. But prophets usually lack sufficient knowledge of technology's inner dynamisms to avoid falling into mere extrinsic critiques of technology, let alone to make practical recommendations for "humanizing" it. A more serious lack is an ethics which operates as a "means of the means" and is rooted in critical reflection on the value content of social action and concrete policies.[9] Ethics as a "means of the means" gets inside the dynamism of any instrument and bends that dynamism to the service of desired values. It avoids mere moralizing about technology as well as simple technological "fixes" which but reinforce the technological imperative. Contemporary societies will master technology only if they are willing to forego specific technologies or their "benefits" when these obstruct more essential values. What is implied is not giving up all technology but combating the technological vision of efficiency. Paradoxically, modern technology can be controlled only by individuals and societies which dethrone technology as their primary source of values. They must *will* to adhere to notions of rationality, efficiency, and problem-solving which "put technology in its place."

What specific value constellations should be placed in command over the technological processes? Some societies list these values as the defense of their cultural integrity, the achievement of institutional reciprocity in dealing with others, and obtaining decent material sufficiency for their people. Dissident countercultures in rich lands lay great emphasis on the manageable scale of operations, psychological satisfactions in work,[10] and simple communitarian living. Socialist revolutionaries, in turn, stress the creation of revolutionary consciousness as a prelude to building "the new man." Because ideology is itself a major source of social values, a sharp break must be made with the "technological ideology" if social mastery of technology is to become possible. "Technological ideology" renders technology normative of all perceptions of social reality. Consequently, whenever "technology assessment" fails to posit valid counternorms for perception, it disqualifies itself as an instrument of responsible social planning. In its broadest sense technology assessment is

> the thorough and balanced analysis of all significant primary, secondary, indirect and related consequences of impacts, present and foreseen, of a technological innovation on society, the environment or the economy.[11]

Although popular images of technology assessment stress its analysis of adverse effects, it is equally concerned with expected benefits. Assessment is a mechanism designed not to halt the advance of technology but rather to determine whether a given technology should be employed. This procedure is best conceived

> as a tool of technology management, as a necessary link between research and development and the needs of society.[12]

The essential problem is not technology itself but the successful management of it, which requires wisdom and clarity as to the kind of society desired and the ways in which technology can help construct such a society. Technology assessors examine possible alternatives; they anticipate and weigh probable effects of technologies on such domains as employment, ecological health, urban concentration, alienation in work situations, distribution of benefits, and transformation of specific behavior; and they generally aim at making "wise" decisons regarding technology. A battery of diagnostic instruments is used, with most technology assessors favoring a systems approach in their efforts "to find the optimal way of briefing the decision-maker (a politician, a manager, or the public)."[13] The unanswered question underlying all efforts at evaluating technological systems bears on the soundness of what Arnstein and Christakis call "the assessors' research paradigm."[14] These authors reject the "epistemological ethnocentrism" operative in Western systems thinking. To "the logic of opposition" they oppose an alternative logic based on mutualism, evoking favorably the multi-element mutualism of the Navajos, the complementarism of the Chinese, and the contextualism of the Japanese. Arstein and Christakis acknowledge that diverse epistemologies determine how technology assessment is put to use and conclude that technology assessment itself, like any other technique, is both a promoter and a destroyer of values.

Special difficulties attend the application of technology assessment. All highly specialized operations are stalked by the first danger, namely, losing touch with the real world. Specialists who juggle models, scenarios, and other abstractions easily lose their sense of what is real and what is not. Only by great effort can technology assessors maintain lines of communication with the people who are the alleged beneficiaries of their decisions. One experiment to link up experts and the general populace is described by Krauch as follows:

> This simulation was run three times in differing modes. First, we instructed the role-players to create an ordinary rational debate. The second mode was a little more sophisticated, more pseudodynamic; it was an ordinary polite debate. In the third case, we instructed the role-players to fight and to try to smash the underlying assumptions of their opponents. The results were quite striking. The judges said that only the third approach really enlightened them and enabled them to make a decision.[15]

Experimental refinements took the form of submitting the working assumptions adopted by experts before reaching their decisions to the judgment of representative citizens. Initial evidence suggests that one can make technical experts accountable to a general public.[16] These probes constitute a seminal effort to harness technology itself (in the form of telephones, computer consoles, and data banks) to the task of involving the public in technology decisions. Indeed all members of society should assert what Borremans and Illich call "political control of the technological characteristics of industrial products."[17] Political communities ought to debate the technological ceiling under which they choose to live. Only thus will "expert" decisions avoid the twin evils of manipulative elitism and technological determinism. Indeed, to place technology assessment under public controls may be the best way to subordinate technology to other values.

Hetman thinks that "if social aspirations are to orient technology in new directions social goals must be stated in terms of objectives and feasible tasks. This can be done only through a truly participatory exchange of ethical and political principles and aspirations.[18] Obviously the general public needs a scientific basis to inform its political decisions. But any scientific basis for decisions is too narrow and one-sided when applied by experts alone. Moreover, scientific rationality will suffer from the same debilities when it is extended to larger numbers of people. Therefore, a "wisdom to match our sciences" must be created. Whence will such wisdom come? Have not science and technology led their adepts away from paths of wisdom, that special unity achieved only after crossing diversity? Many pretechnological societies doubtless generated certain forms of wisdom. Through language and symbols these societies initiated members to a synthesis of all the experience, direct and vicarious, which fell within their ken. This synthesis, expressed in festivities and rituals, brought to daily existence a sense of mystery, of transcendence, even of gratuity—the spontaneous summons to cherish life and beauty for their own sakes. More importantly, ancient wisdoms conferred patterns of meaning to birth, to daily routine, to change, to suffering, and even to death itself. Unfortunately, these wisdoms were imperfect and fragile and suffered from three defects: they were provincial, static, and naive. The present technological age, however, is characterized by traits directly opposed to ethnocentric parochialism, to fixity, and to naiveté. As a result, uncritical wisdoms quickly grow obsolete and crumble under the onslaught of modern science and technology. By revolutionizing humanity's reflective consciousness in three crucial domains, Darwin, Marx, and Freud have in effect buried ancient wisdoms. Darwin made it impossible to view nature ever again as a static system: evolutionary process is its very "essence." And thanks to Marx, history can never again be viewed as linear progress or as cyclic repetition; it is a conflictual process rooted in competing interests. To these demystifications Freud added a third—the demonstration that the overt intentions of human agents habitually mask, in unconscious realms, profound self-delusions. How then can any wisdom be "functional" nowadays if it is static, ahistorical, or ingenuous? Whatever values they may still retain, ancient wisdoms must confront the challenges posed by modern consciousness, itself so powerfully reflected in technology.

Equally intractable problems face technologically "advanced" societies. For the most part these societies have abdicated the very quest after wisdom; their analytical triumphs are paid in the coin of an atrophied ability to grasp totality. Worse still, the fetishistic worship of empirical verification has blinded these societies to the depths of being and meaning beneath surface realities. Modern societies glibly substitute verification for truth and embrace narrow forms of rationality which leave no room for gratuity, for value criteria to govern choices, or for wisdom itself.

Thanks to its seemingly boundless power to dominate nature and satisfy humanity's material wants, technology poses mighty challenges to ancient wisdoms: It raises troubling questions about the ultimate sources of knowledge and power. Nevertheless, modern critics acknowledge that science itself needs to be informed by a new wisdom, some architectonic vision of holistic meanings. But never again can holistic structures of meaning be framed in dogmatic or ethnocentric terms. Hence a wisdom for our times calls for numerous creative dialogues in discourse and in social praxis (critical reflection allied to reflective practice).[19] Such exchanges will fail unless genuine reciprocity presides over them: "old" and "new" mentalities must talk as equals. Yet reciprocity in cultural dialogue can be achieved only if prevailing patterns of economic, social, and political domination are eliminated. More specifically, scientific "experts" must come to acknowledge that they are not expert in the domain of dialogue. Nonetheless, they would be derelict if they did not radically challenge traditional wisdoms as to their assumptions regarding nature and human possibilities. In turn, these wisdoms will need to criticize the value premises of the scientific vision. Neither party to the discourse can do without the other. At stake lies the answer to the perennial question: Can human beings create their own history? Or are citizens of all societies condemned to remain mere objects of history, tossed about by social, political, and conjunctural forces they cannot control?

Social planners and futurists in growing numbers now reflect on the philosophical dimensions of that ambitious enterprise called "managing technology." Arthur Harkins, invoking the work of Maruyama, approves his conclusion that "epistemological or logical 'resonance' is required before assessment/prediction/implementation of cultural phenomena becomes consensual."[20] Harkins warns "experts" against determinism and urges them not to forget that new models emanating from the creativity of individuals can break the bonds of existing systems. Because all previous societies, he adds, have viewed themselves as complex, present complexity does not argue in favor of unmanageability. He endorses the view that one important source of "informed, collective wisdom in developing and managing social/cultural/personal inventions" will be participatory democracy.[21] The same need for philosophical dialogue is affirmed by Mitroff and Turoff, who take it as axiomatic that systems engineers cannot dispense with philosophy if they are to relate their forecasting work to the needs of societies. Normative criteria are indispensable if for no other reason than that identical data utilized by forecasters of different ideological, cultural, or methodological persuasions can be made to support any number of theoretical models. In a word, "data are *not* information; information results from the interpretation of data."[22] Not only does one searching for valid interpretation need "serious and ethical considerations," but "the process of studying the future becomes inseparable from the process of studying the past. A good forecaster should therefore be a good historian."[23] These scientists conclude that "it is the philosophical ability to be self-reflective that separates science from mythology."[24]

Ecologically minded futurists call for "new values suited to spaceship earth." The great task is to link piety toward nature and piety toward all of one's fellow human beings: to marry healthy survival strategies to those which favor justice and equity on planet earth. The quest for a new wisdom to manage technology reflects the judgment of historian Christopher Dawson that

> the true makers of history are not to be found on the surface of events among the successful politicians or the successful revolutionaries: these are the servants of events. Their masters are the spiritual men whom the world knows not, the unregarded agents of the creative action of the Spirit.[25]

Dawson's point is that a new developmental wisdom will neither renounce politics nor condemn revolution but will infuse both with spiritual vision. Ethics alone, he insists, cannot generate the desired wisdom, which can be born only in the deeper waters of spirit and the ultimate recesses of humanity. But although ethics is not sufficient, it is necessary. Unless ethics as a "means of the means" can be incorporated organically into the dynamics of technological management, wisdom will lack its minimum infrastructure. Lacking ethics, the best one can hope for is mechanistic problem-solving elevated to the pseudo dignity of sophistication thanks to the use of electronic computers or elegant input models. Notwithstanding the claims of some enlightened futurists, most technology assessment suffers from just such mechanicity. Few futurists grasp the need to build their models in function of that special unity called *wisdom* that comes only after crossing complexity. This unity, always painfully and precariously achieved, is an indispensable antidote to the modern expert's problem-solving hubris and infatuation with fads. Therefore, scientists and technicians who seek wisdom in their efforts to manage future technology will need to be initiated into desiring and accepting "traditional" and "nonscientific" subjective values like personal suffering and indifference to fame. Unfortunately, these attitudes have now been banished from the roster of "modern" virtues—or, at best, relegated to the inner sanctum of private living. Accordingly, pretechnological societies still have much to teach modern societies regarding the importance of rendering socially respectable such attitudes and disciplines as silence, solitude, contemplation, communion with the rhythms of nature, and respect for the dignity of the cosmos. Without these disciplines, no society and no group of social planners can liberate itself from that worship of technique which prevents it from harnessing technique to human ends.[26]

"Can technology be controlled?" The only answer to this question is: Yes, if.... *If* multiple new dialogues among disciplines, cultures, and strata of population are effectively launched. *If* praxis by decision-makers overcomes elite class barriers and answers the deepest aspirations of the populace. *If* moderns discover a wisdom to match their sciences. *If* traditionals revitalize their ancient wisdoms in the face of the challenges posed by modernity. And *if* a new alliance between political and mystical messianism is effectuated.[27]

Control over technology is so vital precisely because it is not perceived as possible given the instruments human societies presently have at their disposal. This paralysis is reflected in the division of expert opinion over the question of whether any continuity can be found between the structure of a technology and its effects. For William Kuhns the theme of determinism

> is the single most important question raised by the new environments...all of them [thinkers discussed in his book] suggest some dimension of technology where control is impossible or futile.[28]

Kuhns lists three schools of thought on the issue of technology and determinism. Mumford, Giedion, Ellul, and Wiener belong to what he calls the "Encroachment of the Machine" school; Innis and McLuhan, to the "Media Dictates Culture" school; and Fuller, to the "Technology Breeds Utopia" school. The latter intrigues because

> Fuller's implied slogan, "Technology Breeds Utopia," means that we have nothing to fear from technology but that anachronistic response, fear itself. Fuller is so sanguine that his determinism hardly appears to be a determinism at all, but a promise of technological cornucopia.[29]

Yet even Buckminster Fuller in his optimism cannot halt inquiry or exorcise fear. There are, as French philosopher Pierre Ducassé notes, good reasons why all thinking humans now fear the loss of the human possibility of critical thought.[30] Philosophers understand the soporific effects arising from the failure to recognize what Søren Kierkegaard termed the "sickness unto death."[31] For this nineteenth-century Danish existentialist, despair is the sickness unto death, and the most tragic state of the disease is that lack of inward alertness which prevents most people from even acknowledging their state. To lack the "riches of inwardness," he writes, "is like squandering money upon luxuries and dispensing with necessities, or, as the proverb says, like selling one's breeches to buy a wig. *But an age without passion has no values, and everything is transformed into representational ideas*."[32]

Fuller's utopian optimism notwithstanding, humanity needs the warnings of Mumford and Ellul that technology transforms everything into representations. Technology is the vital arena where cultures and subcultures will either survive or be crushed; here their absorptive capacity will be tested. The ultimate challenge posed by technological determinism is to culture itself. Is only one culture possible in the future—a technological culture? Or is technology the death of culture, the very antithesis of civilization?

Technology Versus Civilization

Normative consensus over how to deal with change is a vital element in every culture. The term *culture*, as here employed, embraces the way of life of all human groups. It includes all the standardized learning and forms of behavior which others in one's group learn to recognize and expect: language and symbols; multiple forms of organization (family, kin, occupational roles, legitimacy and authority structures, etcetera); heritage (religious, esthetic, ethical, natural). A civilization, in turn, is simply one species in the genus culture, namely,

> that kind of culture which includes the use of writing, the presence of cities and of wide political organization and the development of occupational specialization.[33]

Central to the notion of all cultures are collective identity, boundaries of inclusion or exclusion of individuals (whether based on criteria of space, lineage, or blood), continuity, and a common historical experience. To all these traits must be added a shared sense of responsibility for the maintenance, dignity, and freedom of the group. Technology poses a unique challenge to culture because its own value dynamics run counter to the limits posed by cultural identity, by spatial or territorial loyalties, or by consensual norms of thought and symbolization. The progressive unification of the globe has occurred within a Western framework, but Toynbee believes that "the present Western ascendency in the world is certain not to last."[34] British economist John White explains why:

> By all historical parallels, development in the so-called Third World ought to take the form of the rise of new and competing cultures to contend with the old and dying civilization which is co-terminous with the white western world stretching from California to the Urals. The obvious candidates are in Asia, especially in East Asia, where two societies have succeeded in modernizing on the basis of models of social organization which are historically specific and owe little to the international development industry. Yet two new factors cast doubt on the rele-

> vance of the Toynbee-esque model of the challenge and response of competing cultures:
>
> (1) technology;
>
> (2) telecommunications.
>
> These factors open the anti-developmental and rather depressing possibility of a single and unchallengeable global culture. Can there ever again be a new civilization?
>
> The assumption that development is a generalisable concept must be seen in this context. It is far more potent than the crude instruments of 'neo-colonialism.' It is the last and brilliant effort of the white northern world to maintain its cultural dominance in perpetuity, against history, by the pretence that there is no alternative.[35]

Is there truly no alternative to standardized technology? Is advanced industrial society incorrigibly one-dimensional? Notwithstanding its enchantment with modern technology, will the Third World be lured by technology into betraying its deeper values as fully as has the West its own? The very impact of Western technology on other civilizations has helped non-Western peoples re-educate themselves. Out of the clash of values has come the clear lesson that no single nation or people can forever be the center of the universe. And though the West has spread the virus of acquisitiveness and the idolatry of material success everywhere, almost nowhere has the West won the hearts of other peoples. Even those who grasp after the West's tools or material rewards do not hold the West's culture in high esteem. A historical parallel is worth citing here. When Napoleon conquered Egypt, the Muslim historian Al-Gabarti displayed no interest in the Frenchman's technology or material wares.

> Al-Gabarti showed a nicer discrimination. French technology hit him in the eye, but he persisted in waiting for a sign. For him, the touchstone of Western civilization, as of his own, was not technology but justice. This Cairene scholar has apprehended the heart of the matter, the issue which the West has still to fight out within itself.[36]

Toynbee views Western technology as a kind of scaffolding around which all societies are building themselves into a unified world. Yet this Western-built scaffolding is not itself durable:

> The most obvious ingredient in it is technology, and man cannot live by technology alone. In the fullness of time, when the ecumenical house of many mansions stands firmly on its own foundations and the temporary Western scaffolding falls away—as I have no doubt that it will—I believe it will become manifest that the foundations are firm at last because they have been carried down to the bedrock of religion.[37]

The Al-Gabartis of today's Third World no longer seek a sign of justice before adopting the "developed" world's technology; they are wise enough to know that this particular sign will not appear. Nevertheless, they intuitively understand that technology can outlive the "civilization" that diffuses it. Frequently, their vision is more lucid than that of Westerners whose complacency over their technological triumphs blinds them both to the injustices they commit in spreading the *imperium* of technology and to the value impasses the West has created for itself.[38]

Technology now threatens to annihilate the human species, to destroy the planet's capacity to support life, and to eliminate human meanings in life. Small wonder, then, that Innuits (Eskimos)—prototypes of a pretechnological people living at a rudimentary cultural level—deem themselves superior to technologically advanced counterparts. Given the sketchiest training, Innuits master tractors and bulldozers better than the Kabloona—the White Men. They quickly learn how to maintain and repair all types of machinery, and no visitor can ever learn as much as they already know about Arctic conditions. As Lord Ritchie-Calder reports:

> That is why they call the Eskimo *Innuit*, the Real Man. They know that *Kabloona* cannot exist in Eskimo country without a welter of civilized equipment such as heated houses, radios, aircraft, supply ships, and so on, while everything an Eskimo family needs to sustain life under the harshest conditions can be carried on a single dog sledge. When *Kabloona* goes traveling by land it is *Innuit* who must show him the way. So, since he can learn White Man's ways quicker than the White Man can learn his, the Eskimo, without arrogance, knows that he is the Real Man.[39]

Like the Innuits, other Third World culture groups may prove able to master Kabloona's technology more quickly than the White Man can learn Innuit's independence or flexibility. Perhaps only societies which for centuries have respected nature can adapt technology in a non-Promethean mode. Can it be that only cultures which cherish community and kin relationships have long-range survival capacities in a world where competition will prove to be not only socially rapacious but dysfunctional to survival as well? "Conciliatory" speeches from First World leaders purvey a "trickle down" imagery: the rich are to get still richer but, in the process, something will be left over for the poor to improve their lot.[40] This view is hardly calculated to induce, in arenas of global development, a "wisdom to match our sciences." On the contrary, it exacerbates the very inequalities which technology breeds and which in turn reinforce technology's own tendency to become a self-validating end.

In international discussions, "developed" countries display a terminological schizophrenia parallel to the one they employ domestically. The French political theorist Raymond Aron contends that

> industrial societies proclaim an egalitarian conception of society; yet at the same time they give rise to collective organizations which are increasingly gigantic and to which individuals are pro-

> gressively more integrated. They spread an egalitarian conception but create hierarchical structures. Thus every industrial society needs an ideology to fill the gap between what men live and what, according to ideas, they ought to live. We observe an extreme form of this contradiction in Soviet society where, in the name of an ideology of abundance, consumption is curtailed as much as possible in order to increase the power of the collectivity. And the American ideology which allows the reconciliation of hierarchic structure with the egalitarian ideal is the ancient formula: "Every infantryman carries in his knapsack a field marshal's baton."[41]

Dichotomies between rhetoric and reality flow necessarily from technology's character as simultaneous bearer and destroyer of values. Technologies of persuasion and image-making "transform culture into luxury"[42] and atrophy the capacity to innovate. Technical integration so totally absorbs even revolution that "the supreme luxury of the technical society will be to grant the bonus of useless revolt and of an acquiescent smile."[43] Scott Buchanan sees Ellul's warning as a summons

> to recover our truly scientific understandings, our objective knowledge of our ends and the ends of nature, and our individual and common wills. This might give us back our reverence and love of nature as well as our shrewd ingenuities in exploiting it.[44]

Optimism with respect to developed countries seems unfounded, however, for even in times of crisis they seem unable to demystify technology. As a result, many observers place their hopes in the Third World. The Palestinian physicist A.B. Zahlan observes that

> these undeveloped human cultural entities may be structures within which fresh and non-Western relationships between science, technology and man appear that may help resolve the numerous diseases of Western society. In other words, it is in the very interest of Western society and the human race to restrain their cultural imperialism and/or to find measures to promote native creativity in Third World countries.[45]

Indeed the very inability of some poor nations to achieve "development" may prove a blessing in disguise, enabling them to avoid that economic "cannibalism" by which nations devour their own prosperity.[46]

Technological idolatry confirms in societies alienating forms of development. This is no argument for rejecting technology, although technological optimists tend to brand any critique of technology as intellectual Ludditism. Criticism, however, is a plea for cultural wisdom to guide technology. And, as E.F. Schumacher writes,

> wisdom demands a new orientation of science and technology towards the organic, the gentle, the non-violent, the elegant and

> beautiful. . . . We must look for a revolution in technology to give us inventions and machines which reverse the destructive trends now threatening us all.[47]

Theorists of social change speak of "viable" and "unviable" nations, warning us that many extant cultures may prove unable to assimilate technology without "losing their soul." Ironically, however, today's technologically "advanced" societies may well be the first to fall victim to generalized anomie, to which they have rendered themselves vulnerable by their pursuit of gigantic size, their compulsive voracity to consume, and their impotence in rewarding creativity except in modes which reinforce technology's sway. The collapse of the industrial world would not surprise Toynbee, however; one recurring theme in his *Study of History* is the existence of an inverse relationship between the cultural level of societies and their degree of technological attainments.[48] Given that any human group's psychic energy is limited, if it channels most of it to solve technological problems, little is left for truly civilizational creativity in esthetic and spiritual domains. The price paid for success in science and technology is often regression on more important fronts, a societal analogue of the tragic persona familiar to our age: the brilliant scientist or industrialist who is emotionally a child and politically an idiot. Toynbee writes that

> man's intellectual and technological achievements have been important to him, not in themselves, but only in so far as they have forced him to face, and grapple with, moral issues which otherwise he might have managed to go on shirking. Modern Science has thus raised moral issues of profound importance, but it has not, and could not have, made any contribution towards solving them. The most important questions that Man must answer are questions on which Science has nothing to say.[49]

The "developed" West may be obliged to return to a hierarchy of values like that which characterized China during the "Middle Ages." Harvard's Everett Mendelsohn, an historian of science, thinks that

> had a visitor from Mars dropped down then, roughly any time from the 5th Century B.C. to the 15th Century A.D., Europe would have seemed the least likely place for the technological revolution to occur. . . for technique to be introduced as *the* rationale of human activity. China, I would guess, would have seemed a much likelier place. Its technology was far more developed; it had a more rationalized commerce and was a more sophisticated bureaucracy. The mandarins made their counterparts in the Vatican look like peasants in terms of the use of knowledge, of written language, of symbolism, and in terms of *their understanding of the position of technique in human life.*[50]

Modern China has turned its back on Confucianism, but its revolutionaries subordinate technique to politics and values. China's early experience with Western technology taught it the lesson that uncritical acceptance of technology leads ultimately to competition, waste, and exploitation. Because technology has to be subordinated to other values, all societies, "developed" and "underdeveloped" alike, will need to revitalize their traditions to serve their future.[51]

One conclusion reached in the present study is that technology can be controlled if it is not sought as an absolute. Paradoxically, technology is indispensable in struggles against the miseries of underdevelopment and against the peculiar ills of overdevelopment. Technology can serve these noble purposes, however, only in those societies in which ideology, values, and decisional structures repudiate the tendency of technology to impose its own logic in striving after goals. Toynbee hopes for the advent of wisdom from efforts by the world's higher religions—Buddhism, Hinduism, Islam, Judaism, and Christianity—to come to terms with universalism and secularism. Lewis Mumford prefers to remind us that civilizations of the past

> did not regard scientific discovery and technological invention as the sole object of human existence; for I have taken life itself to be the primary phenomenon, and creativity, rather than the 'conquest of nature,' as the ultimate criterion of man's biological and cultural success.[52]

Glorifying life and creativity, however, does not guarantee the fullness of their development. Life also comes to an end, and civilizations too, as Paul Valéry poignantly reminds us, are mortal. And technological creativity can be put to destructive purposes. This danger revives ancient philosophical questions as to the meaning of death, of suffering, of tragedy, of ultimate meaning.[53] All known civilizations have answered these questions in religious, albeit not always in transcendental, terms. Consequently, the religious myth of Prometheus illuminates the destiny of civilizations in a posttechnological age.

If humankind is a despairing Prometheus plagued by guilt over having stolen from heaven the divine fire called technology, it cannot avoid being enslaved by its own creation. If, on the other hand, humankind accepts technology as a free gift of the gods enabling the construction of a better world and a closer affinity with the divine, it remains possible for human beings *not* to fall into the idolatry trap.[54] It is no accident that it is precisely within allegedly "one-dimensional" societies like the United States that the strongest voices are heard warning against the twin evils of antitechnological idiocy and romantic technological optimism. Myron Bloy, a theologian and author of *The Crisis of Technological Change,* sees technology bringing new freedoms and new capacities for basing an emerging culture on critically defined norms and values. He explains that, during the technological era,

> God is, in effect, kicking us in the pants and telling us that it is time to grow up. We are given the tools needed to shape a new culture and allowed to use them effectively only in the service of a prophetic commitment. . . . There is no assurance that society will accept this challenge rather than hide in increasingly frenzied operationalism or increasingly brittle idealisms until we are overwhelmed by chaos, but these are our only two options.[55]

These two options now confront not the United States alone but the entire world. The first choice is prophetic commitment to peace, justice, material sufficiency for all, ecological integrity, and the rebirth of vital cultural diversity.[56] The alternative, inevitable if the first option is declined, is chaos: exploitative development for the few at the expense of the many, war-making, technological servitude, ecological pathology, and the reification of all human values.

This study of value conflicts in technology transfer has attempted to peel away the mystifications which veil the true impact of technology on societies nurturing diverse images of development. Technology is revealed herein as a two-edged sword, simultaneously bearer and destroyer of values. Yet technology is not static: it is a dynamic and expansionist social force which provides a "competitive edge" enabling its possessors to conquer economic, political, and cultural power. Consequently, Third World efforts to harness technology to broader developmental goals are paradigmatic of a still greater task: to create a new world order founded not on elitism, privilege, or force but on effective solidarity in the face of human needs. The gestation of a new world order poses two troubling questions for all societies: Can technology be controlled, and will culture survive?

To these two questions the answer is a qualified yes. But several conditions must first be met. Those who aspire to master technology must learn to look critically and constructively at their own cultural wisdom. This searching look at the past is needed if they are to escape the reductionism which impregnates the technological cast of mind. It is to be hoped that out of the confrontation between past values and present technological necessities may emerge new sources of life, creativity, and organic thinking.

New forms of knowledge must be born. French sociologist Edgar Morin pleads for

> a restructuring of the general shape of knowlege. . . a totally new conception of science itself which will challenge and overturn not only established boundaries among disciplines but the very cornerstones of all paradigms and, in a sense, the scientific institution itself.[57]

Only thus can human knowledge adequately explain "the anthropological trinity of species, society, and the individual."[58]

The revitalization of traditions, values, and wisdoms in the light of modern technological challenges and the construction of new modes of understanding must occur at two levels. While particular loyalties and values are revived, more universal attachments to a global order must also gain sway. World-order thinking is essential, writes Indian economist Rajni Kothari, because

> it is no longer possible to bring about successful change of an enduring kind in one area or country, except in very marginal ways, without taking account of the world context. Even revolutions suffer from this limitation. Similarly, no amount of either pleading or moralizing to restrain standards of consumption or curb 'chauvinist' tendencies is likely to go far in the poorer regions unless at the same time a similar onslaught is directed at the citadels of affluence and the centres of political and military dominance.[59]

New planetary bargains must be struck between the rich and poor, the technologically advanced and those less so.[60]

Can a global order promote just development, technological wisdom, ecological health, and reciprocity among all societies? The options are posited by Reimer in these terms:

> Effective curtailment of world population and of energy and other technological uses will require either a world dictatorship, for which history provides no model, or an ethical social order for which there is even less historic precedent. Failing control by one of these means, the industrial world cannot survive. If the industrial world breaks down, however, only the same alternatives remain as suitable models for a viable new social order. In this case, however, an additional possibility occurs; namely, that no reconstruction but an indefinite period of barbarism might ensue.[61]

The "developed" West has shaped modern technology and aggressively exported it to other societies, most of whom received it avidly. While processes of technology transfer have solved innumerable problems, they have likewise destroyed many of the cultural values societies need to achieve a wisdom to match their sciences. The tragic truth is, as Mumford writes, that

> Western man not merely blighted in some degree every culture that he touched, whether "primitive" or advanced, but he also robbed his own descendants of countless gifts of art and craftmanship, as well as precious knowledge passed on only by word of mouth that disappeared with the dying languages of dying peoples.[62]

Many Third World leaders resignedly accept the destruction of their own cultures in order to gain modernity. A general uneasiness has come to prevail, therefore, in all areas where development is discussed. Visions of brave new worlds are no longer euphoric; even erstwhile champions of development have grown fearful of apocalypse. Especially in the rich world, social critics grow weary and pessimistic and come to fear developmental change. [63] All societies, developed and nondeveloped, are being forced to make what French philosopher J.M. Domenach calls a "return to the tragic."[64] No longer do any certitudes exist regarding the course of technology or the future of humankind. Yet this very obscurity is salutary; our age has learned that easy certitudes are mere tranquilizers peddled in the markets of meaning.

Technology is no panacea for the ills of underdevelopment; even at best its promise is uncertain. And no romantic flight from technology can bring salvation from the alienation specific to "developed" societies. For every historical experience of social change is, as Domenach reminds us, true tragedy "thrusting us to the very heart of those relations which any society has of its own self-image, its language, its history and its future."[65]

As all societies struggle to create a world of genuine development, value conflicts will endure. But these conflicts, like technology itself, can prove beneficial. The key lies in the criteria chosen to decide which values will be destroyed and which will be preserved. Technology is indeed a two-edged sword, at once beneficent and destructive. But so is development itself. So is all of human history.

Notes

Introduction *(text pp. 3-12)*

1. For an introduction to the general literature, see: the numerous *Research Reports* issued by UNITAR (United Nations Institute for Training and Research), New York; *Choice and Adaptation of Technology in Developing Countries: An Overview of Major Policy Issues* (Paris: Development Centre, Organisation for Economic Co-operation and Development, 1974); Pierre F. Gonod, *Clés pour le transfert technologique* (Washington, D.C.: Economic Development Institute, International Bank for Reconstruction and Development, August 1974); J.M. Lester, *Technology Transfer and Developing Countries: A Selected Bibliography* (Washington, D.C.: George Washington University, 1974); Daniel L. Spencer and Alexander Woroniak, eds., *The Transfer of Technology to Developing Countries* (New York: Praeger, 1967); *Technology Transfer: A Selected Bibliography,* rev. ed. (Washington, D.C.: National Aeronautics and Space Administration, 1971); Jack Baranson, *Technology for Underdeveloped Areas: An Annotated Bibliography* (Elmsford, N.Y.: Pergamon Press, 1967); and Sherry R. Arnstein and Alexander N. Christakis, *Perspectives in Technology Assessment* (Jerusalem: Science and Technology Publishers, 1975).

2. See, in particular, Denis Goulet, *Etica del desarrollo* (Montevideo: Instituto de Estudios Políticos para América Latina, 1965); *The Cruel Choice: A New Concept in the Theory of Development* (New York: Atheneum, 1971); *A New Moral Order: Development Ethics and Liberation Theology* (Maryknoll, N.Y.: Orbis Books, 1974); and (with Michael Hudson) *The Myth of Aid: The Hidden Agenda of the Development Reports* (New York: IDOC/North America, 1971).

3. For conflicting views of global solidarity, see Denis Goulet, "World Interdependence: Verbal Smokescreen or New Ethic?" Development Paper no. 21 (Washington, D.C.: Overseas Development Council, 1976).

4. Goulet, *The Cruel Choice,* p. x.

5. This definition is explained in greater detail in Denis Goulet, "The Paradox of Technology Transfer," *Bulletin of the Atomic Scientists,* vol. 31, no. 6 (June 1975), pp. 39-46.

6. Charles Cooper, "The Transfer of Industrial Technology to the Underdeveloped Countries," in *Institute of Development Studies Bulletin* (University of Sussex), 3 October 1970, p. 3.

7. This definition is elaborated further in Denis Goulet, "An Ethical Model for the Study of Values," *Harvard Educational Review,* vol. 41, no. 2 (May 1971), pp. 205-227.

8. From a group statement at the Third International Conference of Ministers of Science of OECD Countries, "The Impact of Science and Technology on Social and Economic Development," OECD [Organisation for Economic Co-operation and Development] *Observer,* no. 33 (April 1968), p. 36.

9. On the difference between a "possible" and a "potential" resource, see L.J. Lebret, *Dynamique concrète du développement* (Paris: Les Editions Ouvrières, 1961), pp. 195ff.

10. This classification is drawn from G.R. Hall and R.E. Johnson, "Transfers of U.S. Aerospace Technology to Japan," in *The Technology Factor in International Trade*, ed. Raymond Vernon (New York: Columbia University Press, 1971), p. 319.

11. This problem, applied to education, transportation, and health, respectively, is analyzed by Ivan Illich in his *Deschooling Society* (New York: Harper & Row, 1971); *Energy and Equity* (New York: Harper & Row, 1974); and *Medical Nemesis: The Expropriation of Health* (New York: Pantheon, 1976).

12. For an illuminating study of the role of emotions, fears, and dreams on corporate decisions, see Michael Maccoby, *The Gamesman: The New Corporate Leaders* (New York: Simon and Schuster, 1977).

13. See John R. Seeley, "Administered Persons: The Engineering of Souls," *Interchange* (Canada), vol. 5, no. 3 (1974), pp. 1-13.

14. W. David Ross, ed., *The Works of Aristotle*, 12 vols. (London: Oxford University Press, 1910-1952), vol. 10, *Politics & Economics* (1921), paragraph 1256b.

15. Ibid., paragraph 1265a.

16. Cheikh Hamidou Kane, *Ambiguous Adventure*, trans. Katherine Woods (London: Heinemann, 1972).

17. Chinua Achebe, *Things Fall Apart* (New York: McDowell, Obolensky, 1959).

18. Benjamin R. Barber, *The Death of Communal Liberty: A History of Freedom in a Swiss Mountain Canton* (Princeton, N.J.: Princeton University Press, 1974).

19. Erich Fromm, ed., *Socialist Humanism: An International Symposium* (New York: Doubleday, 1966), p. ix.

20. This phenomenon is studied under the name of the "confidence mechanism" in Charles Elliott, *Patterns of Poverty in the Third World* (New York: Praeger, 1975).

Part One Introduction *(text pp. 15-16)*

1. See Lewis Mumford, *The Myth of the Machine*, 2 vols. (New York: Harcourt Brace Jovanovich, 1971); W.H.G. Armytage, *A Social History of Engineering* (London: Faber & Faber, 1970); D.S.L. Cardwell, *Turning Points in Western Technology* (New York: Neale Watson Academic Publications, 1972); Peter Mathias, ed., *Science and Society 1600-1900* (Cambridge: Cambridge University Press, 1972); John U. Nef, *The Conquest of the Material World* (Chicago: University of Chicago Press, 1964); Charles Singer et al., eds., *A History of Technology*, 5 vols. (London: Oxford University Press, 1958).

2. The term *existence rationality* is explained in Denis Goulet and Marco Walshok, "Values Among Underdeveloped Marginals: Illustrative Notes on Spanish Gypsies," *Comparative Studies in Society and History*, vol. 13, no. 4 (October 1971), pp. 451-472. Cf. also Goulet, *The Cruel Choice*, pp. 188-212.

3. John D. Montgomery, *Technology and Civic Life: Making and Implementing Development Decisions* (Cambridge, Mass.: MIT Press, 1974), pp. 1 (cf. pp. 16-17) and 52.

Chapter 1 *(text pp. 17-30)*

1. On the phenomenological method, see Marvin Farber, *The Aims of Phenomenology* (New York: Harper Torchbooks, 1966); Tran Duc Thao, *Phenomenology and Dialectical Materialism* (Hingham, Mass.: D. Reidel Publishing Co., forthcoming 1978). For specific application of this method to social problems, see Alfred Schutz, *On Phenomenology and Social Relations* (Chicago: University of Chicago Press, 1970).

2. See Dennis Meredith, "Western Technology Abroad," *Technology Review,* vol. 77, no. 5 (March/April 1975), pp. 53-54.

3. It is obvious that technology is not specifically Western and that even so-called "traditional" societies are not static. Cf. Lloyd and Suzanne Rudolph, *The Modernity of Tradition: Political Development in India* (Chicago: University of Chicago Press, 1970) and Mirrit Boutros Ghali, *Tradition for the Future* (Oxford: The Alden Press, 1972).

4. The writer spent six months living with these tribes in 1958. On the values of nomadic populations throughout the world, see Sixten Haraldson, "Socio-Medical Problems of Nomad Peoples," in *The Theory and Practice of Public Health,* ed. W. Hobson (London: Oxford University Press, 1975), pp. 531-542.

5. See Paulo Freire, *Pedagogy of the Oppressed* (New York: Herder and Herder, 1970); *Cultural Action for Freedom* (Cambridge, Mass.: Center for the Study of Development and Social Change and Harvard Educational Review, 1970); and *Education for Critical Consciousness* (New York: The Seabury Press, 1973).

6. On this see Robert Vachon, "Développement et libération dans une perspective interculturelle et cosmique," *Bulletin Monchanin* (Montreal), vol. 8, no. 2, cahier 49 (March/April/May 1975), pp. 3-29.

7. See David C. McClelland and David C. Winter, *Motivating Economic Achievement* (New York: The Free Press, 1969).

8. Danilo Dolci, "Mafia-Client Politics," *Saturday Review,* 6 July 1968, p. 51.

9. Cf. Peter L. Berger, *Pyramids of Sacrifice: Political Ethics and Social Change* (New York: Basic Books, 1974), pp. 166-189.

10. See Marcel Mauss, *The Gift: Forms and Functions of Exchange in Archaic Societies* (New York: Norton, 1967); Wilton S. Dillon, *Gifts and Nations: The Obligations to Give, Receive and Repay* (The Hague: Mouton, 1968); François Perroux, *Economie et société, contrainte, échange, don* (Paris: Presses universitaires de France, 1963); Richard Titmuss, *The Gift Relationship* (New York: Pantheon, 1971).

11. Jacques Ellul, *The Technological Society* (New York: Knopf, 1965). (The French original appeared in 1954). The argument has recently been revived by the publication, in English, of Ellul's *The New Demons* (New York: The Seabury Press, 1975). A negative appraisal of Ellul's view of technology is

given by Richard L. Rubenstein in *Psychology Today,* vol. 9, no. 6 (November 1975), pp. 18 and 20.

12. Lord Ritchie-Calder, "Hell Upon Earth" (Presidential address to the Conservation Society annual general meeting, Caxton Hall, England, 23 November 1968), p. 4.

13. See John U. Nef, *War and Human Progress: An Essay on the Rise of Industrial Civilization* (New York: Norton, 1968).

14. For one illuminating early exchange, see Robert Theobald, "The House That Homo Sapiens Built," *The Nation,* 19 October 1964, pp. 249-252; the reply by Ellul in "The Technological Society: A Dialogue," *The Nation,* 24 May 1965, pp. 567-568; and Theobald's retort, ibid., pp. 568-569. Ellul's views are compared with those of Mumford, Giedion, Wiener, Fuller, McLuhan, and Innis by William Kuhns in *The Post-Industrial Prophets: Interpretations of Technology* (New York: Harper & Row, 1971).

15. See Richard J. Barnet, *The Economy of Death* (New York: Atheneum, 1973).

16. Cf. Arthur M. Okun, *Equality and Efficiency: The Big Tradeoff* (Washington, D.C.: The Brookings Institution, 1975).

17. See M. Taghi Farvar and John P. Milton, eds., *The Careless Technology: Ecology and International Development* (New York: Doubleday, 1972). Cf. also Nicolas Jéquier, ed., *Appropriate Technology: Problems and Promises* (Paris: Development Centre, Organisation for Economic Co-operation and Development, 1976).

18. On this see Paulo Freire, *Education for Critical Consciousness* (New York: The Seabury Press, 1973), pp. 91ff.

19. Cf., for instance, William C. Frederick and Mildred S. Myers, "The Hidden Politics of Social Auditing," *Business and Society Review,* no. 11 (Autumn 1974), p. 49. For more extensive treatment of social auditing, cf. Leslie D. Wilcox, Ralph M. Brooks, George M. Beal, and Gerald E. Klonglan, *Social Indicators and Societal Monitoring: An Annotated Bibliography* (Amsterdam: Elsevier Scientific Publishing, 1975); also, Judith Innes de Neufville, *Social Indicators and Public Policy* (Amsterdam: Elsevier Scientific Publishing, 1975).

20. The call has, in fact, been issued. Cf., for example, *Public Understanding of Science and Technology* (Proceedings of a conference held at the University of Leicester by the Science Policy Foundation of London, June 1976).

Chapter 2 *(text pp. 31-47)*

1. Jacques Ellul, *The Technological Society* (New York: Knopf, 1965), p. 3.

2. Bernard Mandeville, *The Fable of the Bees: Or, Private and Publick Benefits* (Oxford: The Clarendon Press, 1924).

3. Lord Ritchie-Calder, "The Role of Modern Science and Technology in the Development of Nations," in United Nations, *Science and Technology: The Role of Modern Science and Technology in the Development of Nations and*

the Need to Strengthen Economic and Technical Scientific Operations Among States (E/5238), 26 January 1973, add. 1, p. 11.

4. On "product life cycles," see Raymond Vernon, *Sovereignty at Bay: The Multinational Spread of U.S. Enterprises* (New York: Basic Books, 1971); Robert B. Stobaugh, "The Product Life Cycle and World Trade Patterns," in *Proceedings of the Third International Conference of the European Chemical Marketing Research Association* (Amsterdam, June 1970) (Reprinted by Marketing Science Institute as a working paper, November 1970); and Louis T. Wells, Jr., ed., *The Product Life Cycle and International Trade* (Cambridge, Mass.: Harvard University Press, 1972).

5. Frederick Knickerbocker, *Oligopolistic Reaction and Multinational Enterprise* (Cambridge, Mass.: Harvard University Press, 1973).

6. Cf. Jan J. Oyevaar, "Technological Change and the Future of Shipping in Developing Countries," *Development Digest,* vol. 9, no. 4 (October 1971), pp. 53-68.

7. Certain goods are neither luxuries nor necessities but play, for mass consumers, a role functionally equivalent to that of luxuries for the wealthy. Many beauty products are of this type, as are sophisticated cameras, electronic appliances etcetera. What is said above of "luxury" goods, therefore, applies by extension to a wide range of other products.

8. On dependency theory, see Frank Bonilla and Robert Girling, eds., *Structures of Dependency* (Palo Alto, Calif.: Stanford University Press, 1973); also J.D. Cockroft, André Gunder Frank, and Dale Johnson, eds., *Dependence and Underdevelopment* (New York: Doubleday, 1972).

9. This kind of expertise is called, by some, *managerial technology*. Cf. Leon J. Rosenberg and Molly K. Hageboeck, "Management Technology and the Developing World," in *Systems Approaches to Developing Countries,* ed. M.A. Cuenod et al. (Pittsburgh: Instrument Society of America, 1973).

10. These distinctions are explained in G.R. Hall and R.E. Johnson, "Transfers of U.S. Aerospace Technology to Japan," in *The Technology Factor in International Trade,* ed. Raymond Vernon (New York: Columbia University Press, 1971), p. 319.

11. See Norman Gall, "Oil and Democracy in Venezuela," part 1: "Sowing the Petroleum," and part 2: "The Marginal Man," *American Universities Field Staff Reports,* East Coast South America series, vol. 17, nos. 1 and 2 (January 1973).

12. The term *steady state* implies neither changelessness nor stagnation, but a constantly renewed equilibrium and the reconstitution of nonrenewable resources at the same rhythm as these are depleted. See Herman E. Daly, "The Steady-State Economy: What, Why, and How," in Dennis Clark Pirages, ed., *The Sustainable Society* (New York: Praeger, 1977), pp. 107-130.

13. "A Philosophy of Growth," *Koppers 1972 Report,* p. 16. Koppers, headquartered in Pittsburgh, is one of the 200 largest US industrial corporations. It owns more than 100 plants in 32 states, Canada, Western Europe, Australia, and Central and South America and conducts marketing opera-

tions in 67 foreign countries. Its main fields of activity are: chemicals, building materials, precision-machined metal products, and engineering construction.

14. One recent work which advocates "organic" growth is Mihajlo Mesarovic and Eduard Pestel, *Mankind at the Turning Point: The Second Report to the Club of Rome* (New York: Dutton/Reader's Digest Press, 1974).

15. Victor Ferkiss, *The Future of Technological Civilization* (New York: Braziller, 1974), p. 7.

16. Valentina Borremans and Ivan Illich, "La necesidad de un techo común: el control social de la tecnología" (Paper prepared for a seminar at the Centro Intercultural de Documentación, Cuernavaca, Mexico, January 1972), mimeographed. Cf. Illich, *Institutional Inversion,* cuaderno no. 1017 (Cuernavaca: Centro Intercultural de Documentación, 1972).

17. See, for example, Robert Davis, Joel Schatz, and Neil Fisher, "Energy and State Government" (Paper prepared for the State of Oregon, 1 July 1973), mimeographed.

18. E.F. Schumacher, "The Work of the Intermediate Technology Development Group in Africa," *International Labour Review,* vol. 106, no. 1 (July 1972), p. 4.

19. For an introduction to the principles and procedures of this group, see "Intermediate Technology for a Stable, More Prosperous World" and "Annual Report of the Intermediate Technology Development Group," both available from the Intermediate Technology Development Group, London. E.F. Schumacher's philosophy is presented in his *Small Is Beautiful: Economics as if People Mattered* (New York: Harper & Row, 1973).

20. See Nicholas Wade, "Karl Hess: Technology with a Human Face," *Science,* vol. 187, no. 4174 (31 January 1975), pp. 332-334.

21. I am speaking here in purely material terms. Ultimately, of course, the problem of the "good life" must be posed in spiritual terms. For stimulating debate on this issue, see John Friedmann, ed., "Working Papers on the Good Society," Development Paper no. 46 (Los Angeles: University of California, School of Architecture and Urban Planning, n.d.).

22. On this see Mario Kamenetsky, "Engineering Education for a New Society," unpublished paper (Buenos Aires, 1972); also John Friedmann, *Retracking America* (New York: Doubleday, 1973).

23. See Paul T.K. Lin, "Development Guided by Values: Comments on China's Road and Its Implications," *International Development Review,* vol. 17, no. 1 (1975), pp. 2-7.

24. See Geneviève Dean, "A Note on the Sources of Technical Innovations in the People's Republic of China," *The Journal of Development Studies,* vol. 9, no. 1 (October 1972), pp. 187-199; Hans Heymann, Jr., "China's Approach to Technology Acquisition: Part III—Summary Observations," Rand Corporation Document R1575 (Santa Monica, Calif.: 1975); Joseph Needham, "History and Human Values: A Chinese Perspective for World Science and Technology" (Paper presented at the Canadian Association of Asian Studies annual conference, Montreal, May 1975); C.H.G. Oldham, "Science

and Technology Policies," in *China's Developmental Experience,* ed. Michael Oksenberg (New York: The Academy of Political Science, 1973); Lloyd G. Reynolds, "China's Economy: A View from the Grass Roots," *Challenge,* vol. 17, no. 1 (March/April 1974), pp. 12-20; Nicole Ganière, *The Process of Industrialization of China: Primary Elements of an Analytical Bibliography,* working document (Paris: Development Centre, Organisation for Economic Co-operation and Development, 1974); and Jon Sigurdson, "Technology and Employment in China," *World Development,* vol. 2, no. 3 (March 1974), pp. 75-84.

25. These remarks are based largely on unpublished writings of Professor Paul T.K. Lin, numerous conversations with him, and several public lectures by him. I take this occasion to thank Paul Lin warmly for his helpful comments.

26. Goulet, *The Cruel Choice.*

Chapter 3 *(text pp. 53-67)*

1. Interview with Jorge Sabato, Buenos Aires, March 1973.

2. For an initiation to this literature, see the following, Sarah Jackson, *Economically Appropriate Technologies for Developing Countries: A Survey,* Occasional Paper no. 3 (Washington, D.C.: Overseas Development Council, 1972); Edward P. Hawthorne, *The Transfer of Technology* (Paris: Organisation for Economic Co-operation and Development, 1971); Pierre F. Gonod, *Clés pour le transfert technologique* (Washington, D.C.: Economic Development Institute, International Bank for Reconstruction and Development, August 1974); Walter A. Chudson and Louis T. Wells, Jr., *The Acquisition of Proprietary Technology by Developing Countries from Multinational Enterprises: A Review of Issues and Policies* (New York: United Nations Institute for Training and Research, 1971). Also Chudson, *The International Transfer of Commercial Technology to Developing Countries* (New York: United Nations Institute for Training and Research, 1971); Daniel L. Spencer and Alexander Woroniak, eds., *The Transfer of Technology to Developing Countries* (New York: Praeger, 1967).

3. Information obtained in an interview with Charles G. Loring, Cabot Corporation, Boston, 19 July 1973.

4. Enid Baird Lovell, *Appraising Foreign Licensing Performance,* Studies in Business Policy no. 128 (New York: National Industrial Conference Board, 1969), pp. 55-56. Cf. also Piero Telesio, "Foreign Licensing Policy in Multinational Enterprises" (Ph.D. diss., Harvard University, Graduate School of Business Administration, 1977).

5. Recent Argentine legislation on licensing illustrates how LDC governments seek such protection. For a discussion of the issues involved, see "Ciencia y técnica," *La Opinión* (Buenos Aires), 22 September 1974, p. 16; Francisco C. Sercovich, "Dependencia tecnológica en la industria argentina," *Desarrollo económico: revista de ciencias sociales* (Buenos Aires), vol. 14, no. 53 (April/June 1974); Alberto Ribas, Pablo Castro, Carlos Messer, Alberto Santos, and Facundo Oro, *Tecnología y independencia económica* (Buenos Aires: Ediciones Silaba, 1974); and, concerning the specific technology transfer law (ley n. 20.794, passed 27 September 1974), see the

bulletin of the Cámara de Diputados de la Nación, Reunión 26 (Buenos Aires), 3 May 1974, pp. 2790-2792.

6. Thomas Allen, "The International Technological Gatekeeper," *Technology Review,* vol. 73, no. 5 (March 1971), pp. 3-9.

7. A detailed model for use in Latin America has been designed under the name *MONTT* (Mecanismo Organizado Nacional para Transferencia de Tecnología). See Pierre F. Gonod, *Clés pour le transfert technologique* (Washington, D.C.: Economic Development Institute, International Bank for Reconstruction and Development, August 1974), pp. 233-264.

8. See Julian M. Sobin, "Pilgrimage to the Canton Fair," *Columbia Journal of World Business,* vol. 7, no. 6 (November/December 1972), pp. 88-91. Also Geneviève Dean, "A Note on the Sources of Technical Innovations in the People's Republic of China," *The Journal of Development Studies,* vol. 9, no. 1 (October 1972), pp. 187-199.

9. John Phipps and Jenelle Matheson, "Basic Data on the Economy of the People's Republic of China," *Overseas Business Reports* (US Department of Commerce), June 1974, pp. 22-23.

10. In private conversations with the author after the participation of both in the seminar on "Choice and Adaptation of Technology in Less-Developed Countries," Development Centre, Organisation for Economic Co-operation and Development, Paris, 7-9 November 1972.

11. Source: *Science and Technology for Development: Proposals for the Second United Nations Development Decade* (New York: United Nations, 1970), p. 23.

12. The question is investigated in Robert Carrillo Ronstadt, *R&D Abroad: The Creation and Evolution of Foreign Research and Development Activities of U.S. Based Multinational Enterprises* (Ann Arbor, Mich.:University Microfilms, 1977). This is a Ph.D. dissertation submitted to the Harvard University School of Business Administration in 1975.

13. Ibid., p. 23.

14. Information obtained during visit and briefing sessions at Bagó Pharmaceutical Company, March/April 1973.

15. See Alberto Araoz and Carlos Martinez Vidal, *Ciencia e industria, un caso argentino,* Estudios sobre el desarrollo científico y tecnológico, n. 19 (Washington, D.C.: Organization of American States, 1974); Jorge A. Sabato and Oscar Wortman, "Informe preliminar: apertura del paqueto tecnológico para la central núclear de Atucha (Argentina), in *Methods of Evaluation of Technology* (Washington, D.C.: Organization of American States, 1974), pp. 29-95; Sabato, "El debate del uranio," *Mercado* (Buenos Aires), 29 March 1973, pp. 39-43; and Sabato, "Atomic Energy in Argentina: A Case History," *World Development,* vol. 1, no. 8 (August 1973), pp. 23-28.

16. See Miguel S. Wionczek, ed., *Latin American Economic Integration*

(New York: Praeger, 1966). The writings of Sagasti, Kaplan, Kamenetsky, Giral, Herrera, Sabato, Leite Lopes, and others go in much the same line.

17. The role of consultant firms in technology transfers to less-developed countries has been insufficiently studied. Interesting pioneer work has been done, however, by Thomas Vietorisz. See, for example, his "Industrial Development Consultancy and Choice of Technology" (Background paper submitted to Working Group on Industrial Development Consultancy and Choice of Technology for meetings held 26-27 March 1973, Organisation for Economic Co-operation and Development, Paris). Cf. Thomas Aitken, *The Multinational Man: The Role of the Manager Abroad* (New York: John Wiley & Sons, 1973).

18. *CATV, Fulfilling the Potential* (Brochure issued by Arthur D. Little, Inc., Cambridge, Mass., n.d.), p. 2.

19. *This is...Business International* (Brochure issued by Business International Corporation, New York, n.d.), p. 1.

20. Information obtained in interviews with Samuel Pisar and his associates in Paris on 5 October 1973. Pisar is the author of *Coexistence and Commerce: Guidelines for Transaction Between East and West* (New York: McGraw-Hill, 1970).

21. Cf. periodic World Bank and IDA news releases summarizing terms of loans and investments. Most projects mention consultants. One typical project states that "a consulting firm will be retained to provide the senior staff of Agency X; a plant-breeding consultant will be employed during the planting period; consulting engineers will draw up specifications and tender documents for the oil mill and also supervise construction; and consultants have been engaged for the technical assistance to the Agricultural Development Bank."

22. Although in recent years US consultant firms have harvested a bounty crop of contracts from dollar-rich oil-exporting countries.

23. For more details, cf. Denis Goulet, "The Suppliers and Purchasers of Technology: A Conflict of Interests," *International Development Review*, vol. 18, no. 3 (1976), pp. 14-20.

Chapter 4 *(text pp. 69-87)*

1. For representative bibliographies, see David Burtis et al., eds., *Multinational Corporation–Nation-State Interaction: An Annotated Bibliography* (Philadelphia: Foreign Policy Research Institute, 1971); United Nations Library, *Multinational Corporations: A Select Bibliography* (ST/LIB/30), 27 August 1973; Organization of American States, Executive Secretariat for Economic and Social Affairs, *Annotated Bibliography of Transnational Enterprises with Emphasis on Latin America,* August 1974. While specialists continue to debate the suitability of such designations as *multinational* or *transnational* corporations, I have elected to use the latter throughout this work.

2. United Nations, Department of Economic and Social Affairs, *The Acquisition of Technology from Multinational Corporations by Developing Countries* (ST/ESA/12), September 1974, p. 1.

3. United Nations, Department of Economic and Social Affairs, *The Impact of Multinational Corporations on Development and on International Relations* (E/5500/rev. 1), 24 May 1974, add. 1, p. 16. Also United Nations, Department of Economic and Social Affairs, *Multinational Corporations in World Development* (ST/ECA/190), 1973, pp. 25-26.

4. José R. Bejarano, "Let the Multinationals Help," *The New York Times,* 5 December 1971.

5. See *Tapia Says: Power to the People, Proposals for a New Constitution* (Port of Spain, Trinidad: Tapia House, 1974), p. 5: "We are in the midst of a gigantic revolt against the effects of North Atlantic technological civilization, against the inability of civil society to harness technology to human ends."

6. Orville L. Freeman, "Multinational Companies and Developing Countries: A Social Contract Approach," *International Development Review,* vol. 16, no. 4 (1974), pp. 17-19.

7. Cf. Raymond Vernon, "Multinational Enterprises in Developing Countries: An Analysis of National Goals and National Policies" (Draft dated 8 March 1974, prepared for the United Nations Industrial Development Organization), p. 9.

8. See Richard J. Barnet and Ronald E. Müller, *Global Reach: The Power of the Multinational Corporations* (New York: Simon and Schuster, 1974). For a more favorable view of TNCs, see Neil H. Jacoby, *Corporate Power and Social Responsibility* (New York: Macmillan, 1973) and Richard N. Farmer, *Benevolent Aggression* (New York: McKay, 1972). Cf. also George W. Ball, ed., *Global Companies: The Political Economy of World Business* (Englewood Cliffs, N.J.: Prentice-Hall, 1975); Anant R. Negandhi and S. Benjamin Prasad, *The Frightening Angels: A Study of U.S. Multinationals in Developing Nations* (Kent, Ohio: Kent State University Press, 1975); and Abdul A. Said and Luiz R. Simmons, eds., *The Sovereigns: Multinational Corporations as World Powers* (Englewood Cliffs, N.J.: Prentice-Hall, 1975).

9. On Brazil's policy see Nuno Fidelino de Figueiredo, *A transferência de tecnologia no desenvolvimento industrial do Brasil* (Rio de Janeiro: IPEA/INPES, 1972); also F. A. Biato, E.A.A. Guimarães, and M.H. Poppe de Figueiredo, *A transferência de tecnologia do Brasil* (Brasília: IPEA, 1973).

10. Interview conducted in Buenos Aires, 27 March 1973. I have respected the informant's request for anonymity.

11. Barnet and Müller, *Global Reach,* p. 136.

12. For a description of one such effort see ABT Associates, Inc., *Urban Development Applications Project: Final Documentation* (Report prepared for the National Aeronautics and Space Administration, August 1971).

13. See Patrick Conley, "Experience Curves as a Planning Tool" (Special

commentary published by the Boston Consulting Group, June 1970). The ideas contained in this paper were explained more fully to the author by Mr. Conley in two interviews conducted 18 March and 18 November 1974.

14. Interview with Professor Louis T. Wells, Jr., Harvard Business School, 16 April 1974.

15. Cf. Constantine V. Vaitsos, "Bargaining and the Distribution of Returns in the Purchase of Technology by Developing Countries," *Institute of Development Studies Bulletin* (University of Sussex), vol. 3, October 1970, pp. 16-23.

16. Cf. Seymour Tilles, "The 'Foreign Subsidiary' Syndrome," *Perspective* (Boston: The Boston Consulting Group, 1968).

17. Cf. Walter A. Chudson, *The International Transfer of Commercial Technology to Developing Countries,* Research Report no. 13 (New York: United Nations Institute for Training and Research, 1971), pp. 41-46.

18. For a representative statement of these aims, see K. Abdallah-Khodja, "Discours d'ouverture" to the National Seminar on the Transfer of Technology, Algiers, 8-12 October 1973.

19. On "appropriate" technology see E.F. Schumacher, *Small Is Beautiful: Economics as if People Mattered* (New York: Harper & Row, 1973); also Schumacher, "Intermediate Technology," *The Center Magazine,* vol. 8, no. 1 (January/February 1975), pp. 43-50; *Choice and Adaptation of Technology in Developing Countries: An Overview of Major Policy Issues* (Paris: Development Centre, Organisation for Economic Co-operation and Development, 1974); and Nicolas Jéquier, ed., *Appropriate Technology: Problems and Promises* (Paris: Development Centre, Organisation for Economic Co-operation and Development, 1976).

20. Louis T. Wells, Jr., "Economic Man and Engineering Man: Choice of Technology in a Low-Wage Country," *Public Policy,* vol. 21, no. 3 (Summer 1973), pp. 319-342.

21. For further treatment of these points see D.L.O. Mendis, "Planning the Industrial Revolution in Sri Lanka," Industry and Technology Series Occasional Paper no. 4 (Paris: Development Centre, Organisation for Economic Co-operation and Development, 1975); and George Macpherson and Dudley Jackson, "Village Technology for Rural Development," *International Labour Review,* vol. 3, no. 2 (February 1975), pp. 97-118.

22. Hans Heymann, Jr., "China's Approach to Technology Acquisition: Part III—Summary Observations," Rand Corporation Document R1575 (Santa Monica, Calif.: 1975); Geneviève Dean, "A Note on the Sources of Technical Innovations in the People's Republic of China," *The Journal of Development Studies,* vol. 9, no. 1 (October 1972), pp. 187-199; and Shigeru Ishikawa, "A Note on the Choice of Technology in China," "*The Journal of Development Studies,* vol. 9, no. 1 (October 1972), pp. 161-186.

23. See Jorge Sabato and Natalio Botana, "La ciencia y la tecnología en el desarrollo de América latina," in *América latina, ciencia y tecnología en el*

desarrollo de la sociedad, ed. Amílcar O. Herrera (Santiago: Editorial Universitaria, 1970), pp. 59-76. And also by Sabato: "Quantity Versus Quality in Scientific Research: (I) The Special Case of Developing Countries," *Impact,* vol. 20 (1970), pp. 183ff.; *El rol de las empresas del sector público en el desarrollo científico tecnológico* (Washington, D.C.: Organization of American States, 1972); *Ciencia, tecnología, desarrollo y dependencia* (San Miguel de Tucumán, Argentina: Universidad Nacional de Tucumán, 1971). For a brief summary of Sabato's work, see "Sabato y Botana: el circuito triangular investigación-gobierno-producción," in Floreal H. Forni and Raul H. Bision, *La relación ciencia-tecnología-producción, algunos modelos de politica tecnológica,* Universidad del Salvador (Buenos Aires), Estudios de la Ciencia Latinoamericana, serie B, ensayo n. 1 (July 1972), pp. 31-35.

24. Jorge Sabato, "Incorporating Science and Technology into National Development" (Public lecture delivered in San Miguel, Argentina, 16 March 1973). The author attended this lecture and discussed its contents with Sabato.

25. See James Greene and Michael G. Duerr, *Intercompany Transactions in the Multinational Firm*, Managing International Business series, no. 6 (New York: National Industrial Conference Board, 1970); also Chudson, *International Transfer*, p. 51; and Constantine Vaitsos, "Bargaining and the Distribution of Returns in the Purchase of Technology by Developing Countries," *Institute of Development Studies Bulletin* (University of Sussex), October 1970.

26. Pugwash Conferences on Science and World Affairs, "Draft Code of Conduct on Transfer of Technology," *World Development*, vol. 2, nos. 4 and 5 (April-May 1974), pp. 77-82, and "Recommendations Regarding Minimum Norms to be Adopted in Transfer of Technology" (Latin American official statement on technology at subgroup 3, Science and Technology Roundtable, Caracas, Venezuela, 7-11 October 1974). The Organisation of Economic Co-operation and Development has also issued a draft of a code of conduct.

27. See Celso Furtado, "The Brazilian 'Model' of Development," in *The Political Economy of Development and Underdevelopment*, ed. Charles K. Wilber (New York: Random House, 1973), pp. 297-306.

28. See Sudhir Sen, *A Richer Harvest* (Maryknoll, N.Y.: Orbis Books, 1974); Hal Sheets, "Big Money in Hunger," *Worldview,* vol. 18, no. 3 (March 1975), pp. 10-15; and Susan Demarco and Susan Sechler, *The Fields Have Turned Brown: Four Essays on World Hunger* (Washington, D.C.: Agribusiness Accountability Project, 1975).

29. For interesting insights based on psychiatric research of corporate managers' dreams, aspirations, and wishes, see Michael Maccoby, *The Gamesman: The New Corporate Leaders* (New York: Simon and Schuster, 1977).

Chapter 5 *(text pp. 89-122)*

1. The following case studies have been submitted to interested parties for review and critique. I have made every effort to correct factual errors. In

several instances respondents have disagreed with me on questions of interpretation. Although I have carefully weighed their arguments, I have at times adhered to my own interpretation and assume full responsibility for doing so.

2. The reader may consult David C. Major, "Investment Criteria and Mathematical Modelling Techniques for Water Resources Planning in Argentina: The MIT-Argentina Project," in *Systems Approaches to Developing Countries,* ed. M.A. Cuenod et al. (Pittsburgh: Instrument Society of America, 1973), pp. 235-239. For Major's broader theoretical approach, see "Impact of Systems Techniques on the Planning Process," *Water Resources Research,* vol. 1, no. 3 (June 1972), pp. 766-768.

3. Major, "Investment Criteria," p. 4.

4. Ibid., p. 3.

5. From September 1972 through early 1975, I interviewed the following persons: David Major, MIT (numerous times); Frank Perkins, MIT (twice); five Argentine trainees—Tomas Facet, Juan Valdes, José Suarez, Marcos Elinger (twice), and Javier Pasuchi; Argentine officials who played key roles in negotiating the agreement or executing it—Guilhermo Cano, Aquilino Velasco, Carlos Gimenez, Juan D'Albagni (twice), Enrique Aisiks (twice), and Jorge Sarabia. I am indebted to all for their time, thoughtful exchanges, and helpful insights into the Rio Colorado case. Although the interpretation presented here is based on conversations with them, on my reading of available documentation, and on personal visits to Rio Colorado sites, I nevertheless assume full and sole responsibility for what is written here.

6. Major, "Investment Criteria," p. 5.

7. Special thanks go to Messrs. Mario Bonomi, John B. Deaderick, Ricardo Gallagher, Roger Hicks, Eugene Moore, William J. Pullen, G.N. Sheetikoff, and Herman Venegas.

8. It is interesting to note, however, that a US-based company engineer voiced the opinion that many officers would see the main emphasis of the company, within the US, on systems technology, the very antithesis of hardware. The company views itself, in its dealings with Latin America, as a provider of "hardware," but this is not the case with the 50% of its business located in the United States.

9. As defined in an internal memorandum dated 14 March 1973 dealing with "The Corporate Product."

10. Interview conducted in Lima, Peru, 3 May 1973. This information was confirmed in a prior interview with the manager's deputy in Santiago on 16 April 1973.

11. Information contained in the pages devoted to this case was obtained from a review of pertinent documents as well as interviews with Brazilian government officials concerned with the project: João Paulo Reis Velloso, José Walter Bautista Vidal, and their colleagues. At Arthur D. Little, I interviewed the following company officers, most of them several times:

James Gavin, William Krebs, Reed Weedon, Benjamin Fogler, Richard Lacroix, Frederick March; Claudio Margueron, Waldo Newcomer, William Reinfeld, Kamal Saad, and others. Some discussions bore directly on the cold-foods system, others on broader company issues. I am grateful to all for their time, their ideas, and their helpful introductions to colleagues.

12. General Emílio Garrastazu Médici was president of Brazil from 1969 to 1974.

13. Arthur D. Little, Inc., *Cold Chain Food Systems Development for Brazil: A Preliminary Assessment*, Document C-74806, August 1972, p. 13. (Henceforth cited as *ADL Report*.)

14. Ivan Illich, *Deschooling Society* (New York: Harper & Row, 1971) and *Medical Nemesis: The Expropriation of Health* (New York: Pantheon, 1976). The same phenomenon is examined in Robert J. Ledogar, *Hungry for Profits: U.S. Food and Drug Multinationals in Latin America* (New York: IDOC/North America, 1975).

15. *ADL Report*, p. 1.

16. Ibid.

17. Ibid., p. 2.

18. Ibid., p. 24.

19. Ibid., p. 25.

20. One ADL official, upon reading this passage, objected that consultants should not be portrayed as endorsing the values of clients. My reply is that whatever be their subjective wishes, *they do endorse values when they recommend measures which reinforce these values*.

21. *ADL Report,* p. 26.

22. Ibid., p. 5.

23. Ibid., p. 18.

24. Ibid., p. 24.

25. Ibid., p. 25.

26. Ibid., p. 12.

27. Ibid., p. 27.

28. Ibid., Appendix A, pp. 2-4. Italics mine.

29. For evaluative profiles of ADL, see Michael Shanks, "A Hive of Ph.D.'s," *The Director* (London), January 1964, and B.J. Loasby, "Impressions of ADL," unpublished report, 1965.

30. *ADL Report,* p. 53.

31. For a brief introduction to issues of international tourism in development, see the following short articles, all appearing in *Development Digest*, vol. 13,

no. 1 (January 1975): UNCTAD [United Nations Conference on Trade and Development] Secretariat, "International Tourism and the Developing Countries," pp. 39-42; George Young, "Tourism: Blessing or Blight?" pp. 43-54; John M. Bryden, "A Critical Assessment of Tourism in the Caribbean," pp. 55-64; and Donald H. Neiwiaroski, "Small Hotels: A Proposal," pp. 65-68. Other pertinent writings include: Louis Turner and John Ash, *The Golden Hordes: International Tourism and the Pleasure Periphery* (London: Constable, 1975); Patrick Rivers, *The Restless Generation: A Crisis in Mobility* (London: Davis-Poynter, 1972); Dean MacCannell, *The Tourist: A New Theory of the Leisure Class* (New York: Schocken, 1976). In a more scholarly vein, cf. Emmanuel de Kadt, "Social and Cultural Aspects of Tourism" (Paper presented at the joint UNESCO/IBRD [International Bank for Reconstruction and Development] seminar on the "Social and Cultural Impacts of Tourism," Washington, D.C., 8-10 December 1976).

32. Cited in cover story on André Malraux, *Time*, 18 July 1955, p. 24.

33. The World Bank, *Tourism, Sector Working Paper* (Washington, D.C., June 1972), p. 19.

34. Kenneth R. Hansen, *Latin American Tourism: Prospects, Problems, and Alternative Approaches* (Washington, D.C.: Inter-American Development Bank, 1974), p. 6.

35. *International Tourism and Tourism Policy in OECD Member Countries* (Paris: Organisation for Economic Co-operation and Development, 1973), pp. 12, 29, 42.

36. See, for example, representative issues of *Tourism International Air-Letter* (London), ed. John Seekings, or *Worldwide Operating Statistics of the Hotel Industry: Annual Report* (New York: Howarth & Howarth, 1974).

37. Hansen, *Latin American Tourism,* pp. 25ff.

38. The World Bank, *Tourism, Sector Working Paper,* p. 13.

39. Ibid.

40. Government of Mexico, Secretariat of Government Properties, "Comprehensive Development Plan for the Metropolitan Region of Acapulco," in *Terms of Reference*, 15 September 1972.

41. Hiller's ideas appear in numerous articles, notes from seminars given on "Alternate Tourism Perspectives" at Florida International University in 1973, and speeches. The reader may consult the following: "Caribbean Tourism and the University," *Caribbean Educational Bulletin*, vol. 1, no. 1 (January 1974), pp. 15-22; also a Hiller interview in Morris D. Rosenberg's article "Caribbean Awakening: Questioning the 'Fantasies' of Tourism," *The Washington Post,* 14 October 1973; and "Tourism as a Concept Not a Word," *The Times of the Americas,* 6 February 1974. Also Herb Hiller, "Some Basic Thoughts About the Effects on Tourism of Changing Values in Receiving Societies," in *Marketing Travel and Tourism* (Proceedings of the Seventh Annual Conference of the Travel Research Association, Boca Raton, Florida, 1976) (Salt Lake City: University of Utah, Bureau of Economic and

Business Research, 1976). Cf. also H. Hiller, "Background Paper," in *Report of a Regional Seminar on Tourism and Its Effects* (Nassau, November 1975) (Georgetown, Guyana: Caribbean Regional Centre for Advanced Studies in Youth Work, 1975). Hiller is presently writing a book tentatively titled "The Tourism Factory." I wish to thank Herb Hiller for supplying me with numerous unpublished working papers (in particular, "An Alternative Approach to Caribbean Tourism," and "Guidelines for the 'New Tourism' . . .") and for granting me several interviews. This specific quotation and those following attributed to him are drawn from these working papers.

42. See Lakshmi Persaul, "Tourism: Good or Evil?" *Caribbean Contact,* January 1974, pp. 18-20; Henry Pelham Burn, "The Tourist Connection," *Development Forum* (United Nations), November 1973, pp. 1-2; Louis A. Perez, Jr., "Aspects of Underdevelopment: Tourism in the West Indies," *Science and Society,* vol. 37, no. 4 (Winter 1973), pp. 473-480; Dante B. Fascell, "Tourism and Development in the Caribbean" (Speech delivered to the Caribbean Travel Association, Aruba, Dutch West Indies, 15 September 1973).

43. Remarks of James F. Mitchell, cited in Jerry Kirshenbaum, "To Hell with Paradise," *Sports Illustrated*, 21 May 1973, p. 50.

44. Max Kaplan, "Society and Leisure," *Bulletin for Sociology of Leisure, Education and Culture* (published by European Center for Leisure and Education), vol. 2 (1972); also Kaplan, *Technology, Human Values and Leisure* (New York: Abingdon Press, 1971). Among Kaplan's patrons is Vigdor Schreibman, author of "A Policy Guide for the Las Colinas Concept of a Leisure-Enrichment Environment" (San Juan, Puerto Rico: n.p., n.d.) and of "Description of the Las Colinas Development, a Leisure-Enrichment Community, Fajardo, Puerto Rico" (n.p., n.d.).

45. Thorstein Veblen, *The Theory of the Leisure Class* (Boston: Houghton Mifflin, 1973); Josef Pieper, *Leisure, the Basis of Culture,* trans. Alexander Dur (New York: Pantheon, 1964); Johan Huizinga, *Homo Ludens: A Study of the Play Element in Culture* (New York: J&J Harper Editions, 1970) (original in German, 1944); and Sebastian de Grazia, *Of Time, Work, and Leisure* (Garden City, N.Y.: Doubleday, 1974).

46. For a good statement on these issues, see Louis A. Perez, Jr., "Aspects of Underdevelopment: Tourism in the West Indies," *Science and Society,* vol. 37, no. 4 (Winter 1973), pp. 473-480.

47. The phrase comes from Colombian journalist Enrique Santos Caldera.

48. The present discussion is based on two interviews with the president of ASTARSA, Eduardo Braun Cantilo, and his chief engineer, Bernardo Rikles, plus a visit to the shipyard and an examination of relevant documentation.

49. For a general treatment of technology in the shipping industry, see Jan J. Oyevaar, "Technological Change and the Future of Shipping in Developing Countries," *Development Digest*, vol. 9, no. 4 (October 1971), pp. 53-68.

50. Sources for information on the Ellicott Machine Corporation are: company documents; interviews with the firm's assistant general sales manager, John Stribling, and with its project engineer, Harold Wolfing, Jr.; and a visit to manufacturing plant and research facilities.

51. Sources for this case study are company documents and interviews with the following Cabot officials: Norton Sloan, Charles G. Loring, Don Harper, Willis Hunt, and William May.

52. On this see *Relatório 1970-73* (Brasília: Ministério da Indústria e do Comércio, Instituto Nacional da Propriedade Indústrial [INPI], n.d.). This is a progress report released by the Institute.

53. The name of the corporation is simply USM, letters which do not now stand for anything in particular. Before it diversified, however, the firm bore the official name United Shoe Machinery.

54. Sources are company documents; interviews with USM's James P. McSherry, vice president for corporate development (15 February 1974), and Walter Abel, vice president for research (15 March 1974); and a visit to the R&D site in Beverly, Massachusetts (15 March 1974).

55. Company production lines now include: computer systems, all manner of plastics and foams, chemicals, rubber and plastics processing machinery, adhesives for construction, and shoe machinery.

56. USM *Annual Report*, 1973, p. 10.

57. Ibid., p. 3.

58. I am grateful to Eduardo Bracamonte, Luis Rojas, and Carol Michaels for facilitating my visit to Tiataco and Huayculi.

59. From an unpublished text by Luis Rojas, "Proposiciónes relatives a la educación como desarrollo de la cultura," dated April 1974. Rojas is director of a coordinating team working directly with the cooperatives and known as ASCODE (Asociación de Comunidades de Campesinos para el Desarrollo). The Quechua cooperatives have also based their practice in part on the theoretical work of Argentine sociologist Rodolfo Kusch. For samples of his work, see "Una lógica de la negación para comprender a América," *Nuevo Mundo* (Buenos Aires), vol. 3, no. 1 (January-June 1972), pp. 170-178; also Kusch, "El miedo de la historia," *Comentario* (Buenos Aires), no. 64 (January-February 1969), pp. 3-11.

60. The problem of scarce firewood is worldwide. Cf. Erik P. Eckholm, "The Deterioration of Mountain Environments," *Science*, vol. 189. no. 4205 (5 September 1975), pp. 764-770; and Eckholm, "Cheaper Than Oil, but More Scarce," *The Washington Post*, 27 July 1975.

61. From the US-based Inter-American Foundation (IAF), which assists Latin American self-help efforts in development. IAF funding goes through a Bolivian intermediary, Caritas de Bolivia, a Roman Catholic social-development agency. On the IAF, cf. *They Know How: An Experiment in Development Assistance* (Washington, D.C.: Inter-American Foundation, 1977).

62. A brief statement of project goals is found in William V. Donaldson, A.F. Fath, H.R. Singleton, and C.R. Turner, "Technology Transfer in Tacoma, Washington—The Totem One Program" (Paper presented to the American Institute of Aeronautics and Astronautics, Urban Technology Conference, San Francisco, 24-26 July 1972). For an outsider's description, see "Aerospace Technology Goes to Town," *Industry Week*, 8 January 1973, pp. 55-58.

Chapter 6 *(text pp. 123-142)*

1. See "Report of the Group of Eminent Persons to Study the Impact of Multinational Corporations on Development and on International Relations," in United Nations, Department of Economic and Social Affairs, *The Impact of Multinational Corporations on Development and on International Relations* (E/5500/rev. 1), 24 May 1974, add. 1, pp. 55-56.

2. For a summary evaluation see *Andean Pact: Definition, Design, and Analysis* (New York: Council of the Americas, 1973). Cf. Robert N. Seidel, *Toward an Andean Common Market for Science and Technology* (Ithaca, N.Y.: Cornell University, Program on Policies for Science and Technology in Developing Nations, 1974).

3. For an introduction to the notion of "development poles," see François Perroux, *L'économie du XXième siècle* (Paris: Presses universitaires de France, 1964); and J.P. Baillargeon et al., "Le rôle des pôles dans le développement," *Développement et civilisations*, no. 5 (January 1961), pp. 31-53; and Alexander Gershenkron, *Economic Backwardness in Historical Perspective: A Book of Essays* (Cambridge, Mass.: Harvard University Press, 1962).

4. Robert Girling, "Dependency, Technology and Development," in *Structures of Dependency*, eds. Frank Bonilla and Robert Girling (Palo Alto, Calif.: Stanford University Press, 1973), p. 46.

5. See Ronald Müller, "Poverty Is the Product," *Foreign Policy,* no. 13 (Winter 1973-74), pp. 71-102, and Müller et al., "An Exchange on Multinationals," *Foreign Policy,* no. 15 (Summer 1974), pp. 83-92. Also Doug Hellinger and Steve Hellinger, "The Job Crisis in Latin America: A Role for Multinational Corporations in Introducing More Labour-Intensive Technologies," *World Development*, vol. 3, no. 6 (June 1974), pp. 399-410.

6. In Brasília, 9 January 1975.

7. See José Walter Bautista Vidal, *Projecto do II plano nacional de desenvolvimento (1975-1979)* (Brasília: Federal Republic of Brazil, September 1974), p. 123.

8. See G. Edward Schuh, *Research on Agricultural Development in Brazil* (New York: The Agricultural Development Council, 1970); Wayne G. Broehl, Jr., *The International Basic Economy Corporation* (Washington, D.C.: National Planning Association, 1968); and M. Yudelman, G. Butler, and R. Banjeri, *Technological Change in Agriculture and Employment* (Paris: Development Centre, Organisation for Economic Co-operation and Development, 1971).

9. The effective contract date was 1 September 1974. A report entitled "First Year Report on Contract NIRT 1 Assistance in Telecommunication Planning" is forthcoming from the Institute for Communication Research at Stanford University.

10. See Celso Furtado, "Analisis del 'modelo' brasileño," in *Dos analisis de la economia latino-americana* (Buenos Aires: Centro Editor de América Latina, 1972), pp. 1-127; Furtado, *O mito do desenvolvimento económico* (Rio de Janeiro: Paz a Terra, 1974); Fernando Henrique Cardoso, *O modelo politico brasilero* (São Paulo: Difusão Europeia do Livro, 1973); CoDoC [Cooperation in Documentation and Communication], *Bibliographical Notes for Understanding the Brazilian Model: Political Repression & Economic Expansion*, Common Catalogue no. 2 (Washington, D.C.: CoDoC, 1974); and Albert Fishlow, *Essays on Brazilian Economic Development* (Berkeley: University of California Press, 1976).

11. *World Bank Atlas: Population, Per Capita Product, and Growth Rates* (Washington, D.C.: The World Bank, 1974), p. 19.

12. Special pleading is discussed in Peter L. Berger, *Pyramids of Sacrifice* (New York: Basic Books, 1974). Cf. E.J. Mishan, *Technology and Growth: The Price We Pay* (New York: Praeger, 1970).

13. The reader is referred to John Rawls, *A Theory of Justice* (Cambridge, Mass.: Harvard University Press, 1971); Morris Ginsberg, *On Justice in Society* (Ithaca, N.Y.: Cornell University Press, 1965); and Robert Nozick, *Anarchy, State, and Utopia* (New York: Basic Books, 1974).

14. In his address to the United Nations Conference on Trade and Development in Santiago, Chile, 14 April 1972, McNamara described Brazil's growth pattern in the following way: "In the last decade Brazil's GNP per capita, in real terms, grew by 2.5% per year, and yet the share of the national income received by the poorest 40% of the population declined from 10% in 1960 to 8% in 1970, whereas the share of the richest 5% grew from 29% to 38% during the same period. In GNP terms, the country did well. The very rich did very well. But throughout the decade the poorest 40% of the population benefited only marginally." (From a reprint of the address published by the International Bank for Reconstruction and Development, Washington, D.C.), p. 4.

15. Müller, "Poverty Is the Product," *Foreign Policy*, no. 13 (Winter 1973-74), p. 81, citing Irma Adelman and Cynthia T. Morris, "An Anatomy of Income Distribution Patterns in Developing Nations: A Summary of Findings," Economic Staff Paper no. 116 (Washington, D.C.: International Bank for Reconstruction and Development, 1971). For a broader treatment of income inequality, see Albert O. Hirschman, "The Changing Tolerance for Income Inequality in the Course of Economic Development," *The Quarterly Journal of Economics*, vol. 87 (November 1973), pp. 544-566.

16. Earlier in this chapter I have alluded to studies by Celso Furtado in support of the hypothesis that technology transfer, as presently conducted, has concentrated income in Brazil. Other Brazilian economists disagree, however, including Edmar L. Bacha, whose paper "Recent Brazilian Eco-

nomic Growth and Some of Its Main Problems" will appear in a forthcoming work tentatively titled "Growth and Income Distribution Models for Brazil," ed. Lance Taylor.

17. I have been unable to locate the phrase in Furtado's published writings but have heard him use it often in private conversation and in seminars given by him, one at Harvard University, 7 December 1972, and the other at the American University in Washington, 10 November 1972.

18. E.F. Schumacher, "Growth—Yes, but who for and how fast?" in *Study Encounter*, vol. 9, no. 4 (Geneva: World Council of Churches, 1973), p. 2.

19. See Louis T. Wells, Jr., "Men and Machines in Indonesia's Light Manufacturing Industries," *Bulletin of Indonesian Economic Studies*, vol. 9, no. 3 (November 1973), pp. 62-72, and Wells, "Don't Overautomate Your Foreign Plant," *Harvard Business Review*, vol. 52, no. 1 (January-February 1974), pp. 111-118.

20. Wells, "Men and Machines," p. 63.

21. Ibid., p. 64.

22. See Louis T. Wells, Jr., "Economic Man and Engineering Man: Choice of Technology in a Low-Wage Country," *Public Policy*, vol. 21, no. 3 (Summer 1973), pp. 319-342.

23. John W. Thomas, "The Choice of Technology in Developing Countries: The Case of Irrigation Tubewells in Bangladesh," quoted with permission from preliminary draft, p. 30. Subsequently published as Harvard Studies in International Affairs no. 32 (Cambridge, Mass.: Harvard University, Center for International Affairs, 1975).

24. Ibid., p. 35.

25. See Wells, "Don't Overautomate," pp. 111-118. Cf. Charles Cooper and Raphael Kaplinsky, *Second-Hand Equipment in a Developing Country* (Geneva: International Labour Office, 1974).

26. John Friedmann and Flora Sullivan, "The Absorption of Labor in the Urban Economy: The Case of Developing Countries," *Economic Development and Cultural Change*, vol. 22, no. 3 (April 1974), pp. 385-413.

27. Ibid., p. 389.

28. Ibid., p. 405.

29. On this see International Labour Office, *The World Employment Programme*, Reports ILC/1/I and ILC/56/IV (Geneva: International Labour Office, 1969). Also three ILO country studies on the employment issues: *Towards Full Employment: A Programme for Colombia* (Geneva: 1970); *Matching Employment Opportunities and Expectations: A Programme of Action for Ceylon,* 2 vols. (Geneva: 1971); and *Employment, Incomes and Equality: A Strategy for Increasing Productive Employment in Kenya*, 2 vols. (Geneva: 1972).

30. Robert Theobald, *The Guaranteed Income: Next Step in Economic Evolution?* (New York: Doubleday, 1966).

31. Robert Theobald, "Communication and Change" (Paper presented at the Network for Global Concern seminar, Spokane, Wash., 12-15 February 1975), p. 5.

32. Ibid.

33. Robert Theobald, "The Second Copernican Revolution" (Address delivered at the Smithsonian Institution, Washington, D.C., 26 April 1974), p. 10.

34. Frances Stewart, "Employment and the Choice of Technique: Two Case Studies in Kenya," to appear in the forthcoming *Readings in Employment in East Africa,* ed. M. Godfrey and D. Ghai. Quoted, with permission, from mimeographed manuscript, p. 1.

35. Ibid., p. 7.

36. Frances Stewart, "Technology and Employment in LDCs," *World Development*, vol. 2, no. 3 (March 1974), p. 37.

37. See chapter 5, "Case 3: Frozen Foods in Brazil."

38. On the effects of media on cultural aspirations and expressions, see Denis Goulet, "The Political Economy of the Image Industry," *Newstatements Journal*, vol. 2, no. 1 (1973), pp. 8-14. Also Herbert I. Schiller, "The Appearance of National Communications Policies: A New Arena for Social Struggle," *Gazette,* vol. 21, no. 2 (1975), pp. 82-94; Schiller, "Madison Avenue Imperialism," *Transaction-Society*, March-April 1971, pp. 52-58; Armand Mattelart, Mabel Piccini, and Michele Mattelart, "Los medios de comunicación de masas," special edition of *Cuadernos de la Realidad Nacional* (Santiago), 2nd ed., no. 3 (March 1970); Alan Wells, *Picture Tube Imperialism? The Impact of U.S. Television on Latin America* (Maryknoll, N.Y.: Orbis Books, 1972); and Rita Cruise O'Brien, "Domination and Dependence in Mass Communications: Implications for the Use of Broadcasting in Developing Countries," *Institute of Development Studies Bulletin* (University of Sussex), vol. 6, no. 4 (March 1975), pp. 85-89.

39. Lewis Mumford, *Technics and Civilization* (New York: Harcourt, Brace & World, 1934), pp. 12-18.

40. George Young, "Tourism: Blessing or Blight?" *Development Digest*, vol. 13, no. 1 (January 1975), p. 49.

41. In his recent study *Working* (New York: Pantheon, 1974), Studs Terkel shows eloquently how central the work *milieu* is to people's sense of identity, value, and utility. See also Leonard Goodwin, *Do the Poor Want to Work?* (Washington, D.C.: The Brookings Institution, 1972), and Charles M. Savage, *Work and Meaning: A Phenomenological Enquiry* (Springfield, Va.: US Department of Commerce, National Technical Information Service, 1973).

42. See *New Concepts and Technologies in Third World Urbanization*

(Proceedings of the Second Annual Spring Colloquium sponsored by the Subcommittee on Comparative Urbanization of the Committee on International and Comparative Studies, Los Angeles, 17-18 May 1974) (Los Angeles: University of California, School of Architecture and Urban Planning, 1974).

43. Concern over this troubling question pervaded the deliberations of an interesting seminar attended by the author and held 29 September-3 October 1975 at Bellagio, Italy, under the auspices of the International Federation of Institutes of Advanced Study (IFIAS) on the theme "Cultural Diversity and Development."

44. These are, principally: Donella H. Meadows et al., *The Limits to Growth: A Report for the Club of Rome's Project on the Predicament of Mankind* (New York: Universe Books, 1972); Mihajlo Mesarovic and Eduard Pestel, *Mankind at the Turning Point: The Second Report to the Club of Rome* (New York: Dutton/Reader's Digest Press, 1974); Jay Forrester, *Industrial Dynamics* (Cambridge, Mass.: MIT Press, 1961); Forrester, *Urban Dynamics* (Cambridge, Mass.: MIT Press, 1969); Forrester, *World Dynamics* (Cambridge, Mass.: MIT Press, 1971).

45. Cf. Amílcar O. Herrera et al., *Catastrophe or New Society? A Latin American World Model* (Ottawa: International Development Research Centre, 1976).

46. See, for example, the statement delivered by the Brazilian representative to Committee II, Twenty-Sixth Session of the UN General Assembly, on item 47 of the agenda, United Nations Conference on the Human Environment, 29 November 1971.

47. Recent examples are: John and Magda C. McHale, *Human Requirements, Supply Levels and Outer Bounds: A Framework for Thinking About the Planetary Bargain,* policy paper, Aspen (Colo.) Institute for Humanistic Studies, 1975; and *The Planetary Bargain: Proposals for a New International Economic Order to Meet Human Needs,* (Report of an International Workshop, Aspen Institute for Humanistic Studies, 7 July-1 August 1975).

48. *Poverty in American Democracy: A Study of Social Power* (Washington, D.C.: United States Catholic Conference, 1974), pp. 133-134. Cf. Richard A. Falk, *This Endangered Planet* (New York: Random House, 1971), p. 23. For one Third World view of the global problem, see Jimoh Omo-Fadaka, "An Alternative to Imperialist Development," *The Ecologist*, vol. 2, no. 6 (June 1972), p. 28.

49. Victor Ferkiss, *The Future of Technological Civilization* (New York: Braziller, 1974), pp. 68-69. Italics mine. Cf. Marshall I. Goldman, *Spoils of Progress: Environmental Pollution in the Soviet Union* (Cambridge, Mass.: MIT Press, 1972).

50. Leo A. Orleans and Richard P. Suttmeier, "The Mao Ethic and Environmental Quality," *Science,* vol. 170, no. 3693 (11 December 1970), pp. 1173-1176.

51. On this see George Lutjens, "The Curious Case of the Puerto Rico Copper Mines," *Engineering and Mining Journal,* February 1971; also David

Ackerman, "Mining and the Environment," *Industrial Puerto Rico,* February-March 1970.

52. W.W.S. Charters, "Nuclear Fusion and Solar Energy," *Science and Australian Technology*, vol. 10, no. 5 (October 1972); Charters, "Alternative Energy Sources," *Melbourne University Magazine* (1974), pp. 71-73. See also William E. Heronemus, "The Case for Solar Energy," *The Center Report*, vol. 3, no. 1 (February 1975), pp. 6-9; Arthur D. Little, Inc., "New Directions for Solar Energy Applications" (Report presented to the Environment Subcommittee of the House Committee on Interior and Insular Affairs, June 1973); and Sandy Eccli et al., *Alternative Sources of Energy: Practical Technology and Philosophy for a Decentralized Society* (New York: The Seabury Press, 1975).

Part Three Introduction *(text pp. 145-146)*

1. Samuel L. Parmar, "The Environment and Growth Debate in Asian Perspective," in *Anticipation* (Report of "An Ecumenical Conference on the Scientific, Technological, and Social Revolutions in Asian Perspective," Kuala Lumpur, Malaysia, 23-29 April 1973) (Geneva: World Council of Churches, 1973), p. 4.

Chapter 7 *(text pp. 147-165)*

1. Marshall Wolfe, "Development: Images, Conceptions, Criteria, Agents, Choices"; "Social and Political Structures: Their Bearing Upon the Practicability and Scope of a Unified Approach to Development Policy"; "National Centre and Local Group or Social Unit: Problems of Communication and Participation," all unpublished papers, n.d., cited with permission of the author.

2. Gunnar Myrdal is generally thought to have originated the term *"soft states.* See his *Asian Drama: An Inquiry Into the Poverty of Nations* (New York: Pantheon, 1968), vol. 1, pp. 66-67, and vol. 2, pp. 895-900. For an explicit application, see Franklin Tugwell, "Modernization, Political Development and the Study of the Future," in *Political Science and the Study of the Future*, ed. Albert Somit (New York: Holt, Rinehart & Winston, 1974), pp. 155-173.

3. Mahbub ul Haq, "The Crisis in Development Strategies," in *The Political Economy of Development and Underdevelopment*, ed. Charles K. Wilber (New York: Random House, 1973), p. 370.

4. On Algeria see William B. Quandt, *Revolution and Political Leadership, Algeria 1954-1968* (Cambridge, Mass.: MIT Press, 1969); David and Marina Ottaway, *Algeria: The Politics of a Socialist Revolution* (Berkeley: University of California Press, 1970); Gérard Chaliand, *L'Algérie indépendente* (Paris: François Maspero, 1972); and Etienne Malarde, *L'Algérie depuis* (Paris: La Table Ronde, 1975).

5. Even the third, pessimistic view concedes that material improvement of the needier masses is required and that many institutions need to be "modernized" in order to perform essential tasks.

6. On this see Nicolas Spulber, "Contrasting Economic Patterns: Chinese and Soviet Development Strategies," *Soviet Studies*, vol. 15, no. 1 (July 1963), p. 13. Cf. John Phipps and Jenelle Matheson, "Basic Data on the Economy of the People's Republic of China," *Overseas Business Reports* (US Department of Commerce), June 1974, pp. 11ff.

7. See Errol G. Rampersad, "Burma's Economy Is Deteriorating," *The New York Times*, 24 January 1972. Also Stanley Johnson, "The Road to Mandalay," *VISTA* (Magazine of the United Nations Association of the USA), vol. 7, no. 3 (November/December 1971), pp. 22-29.

8. For a brief overview, see Frank Bonilla and Robert Girling, eds., *Structures of Dependency* (Palo Alto, Calif.: Stanford University Press, 1973). A critical reading of dependency theory is offered by Robert Packenham in "Latin American Dependency Theories: Strengths and Weaknesses" (Paper prepared for the National Development Colloquium, Center for International Studies, Princeton University, 9 November 1973). Cf. José Luis Imaz, "Adiós a la teoría de la dependencia? una perspectiva desde la Argentina," *Sumário* (Buenos Aires), year 7, no. 28 (October-December 1974), pp. 49-75.

9. See Pablo Gonzalez-Casanova, "Internal Colonialism and National Development," in *Latin American Radicalism*, ed. Irving Louis Horowitz, Josue de Castro, and John Gerassi (New York: Vintage Books, 1969), pp. 118-139.

10. On the "Brazilian model," see Celso Furtado, "Analisis del 'modelo' brasileño," in *Dos analisis de la economia latino-americana* (Buenos Aires: Centro Editor de América Latina, 1972), pp. 1-127; Furtado, *O mito do desenvolvimento económico* (Rio de Janeiro: Paz a Terra, 1974). Also Fernando Henrique Cardoso, *O modelo político brasileiro* (São Paulo: Difusão Europeia de Livro, 1973); and Mario Henrique Simonsen and Roberto de Oliveira Campos, *A nova economia brasileira* (Rio de Janeiro: José Olympio Editora, 1974). Of use to English-language readers is CoDoC [Cooperation in Documentation and Communication], *Bibliographical Notes for Understanding the Brazilian Model: Political Repression & Economic Expansion*, Common Catalogue no. 2 (Washington, D.C.: CoDoC, 1974).

11. Gunnar Myrdal, *The Challenge of World Poverty* (New York: Pantheon, 1970), p. 277. Italics are Myrdal's.

12. For one example, see José Walter Bautista Vidal, *Projecto do II plano nacional de desenvolvimento (1975-79)* (Brasília: Federal Republic of Brazil, September 1974), section on "Integration to the World Economy."

13. See the resolution adopted by the Sixth Special Session of the UN General Assembly, *Declaration on the Establishment of a New International Economic Order*, April 1974.

14. On this see Gustav Ranis, "Some Observations on the Economic Framework for Optimum LDC Utilization of Technology," in *Technology and Economics in International Development* (Report of a seminar sponsored by the Agency for International Development, Washington, D.C., 23 May 1972), p. 36.

15. Address delivered in Santiago, Chile, 14 April 1972.

16. Dudley Seers and Leonard Joy, eds., *Development in a Divided World* (Harmondsworth, England: Penguin, 1971).

17. Mahbub ul Haq, "Developing Country Perspective," in *Technology and Economics in International Development* [see note 14], p. 42.

18. Julius K. Nyerere, *Freedom and Socialism: A Selection of Writings and Speeches* (London: Oxford University Press, 1968), pp. 19-20.

19. Ibid., p. 2.

20. Ibid., p. 11.

21. E.F. Schumacher, *Small Is Beautiful: Economics as if People Mattered* (New York: Harper & Row, 1973), p. 147.

22. See, for instance, Jacques Beaumont, "A Sick Society Is No Model for Export," World Council of Churches *Newsletter,* July 1969, pp. 1-5.

23. On this approach within the United States, see Nicholas Wade, "Karl Hess: Technology with a Human Face," *Science*, vol. 187, no. 4174 (31 January 1975), pp. 332-334.

24. For a brief overview see E.F. Schumacher, "The Work of the Intermediate Technology Development Group in Africa," *International Labour Review*, vol. 106, no. 1 (July 1972), pp. 3-20.

25. See Government of India, Ministry of Industrial Development, Appropriate Technology Cell, *Appropriate Technology for Rapid Economic Growth* (New Delhi, 1972).

26. Henry C. Wallich, "The Future of Capitalism," *Newsweek*, 22 January 1973, p. 62.

27. On the general notion of political marketing, see Charles Hampden-Turner, *From Poverty to Dignity: A Strategy for Poor Americans* (Garden City, N.Y.: Doubleday, 1974). Also Hampden-Turner, "A Proposal for Political Marketing," *Yale Review of Law and Social Action* (Winter 1970), pp. 93-100.

28. Karl Mannheim, *Freedom, Power and Democratic Planning* (London: Routledge & Kegan Paul, 1951), p. 191.

29. Speech by Fidel Castro delivered at Santiago, Cuba, 26 July 1973. Reprinted in English in *Granma* (Havana), 5 August 1973, p. 5.

30. David C. McClelland, *The Achieving Society* (Princeton, N.J.: Van Nostrand, 1961), and McClelland and David C. Winter, *Motivating Economic Achievement* (New York: The Free Press, 1969).

31. See Hollis Chenery et al., *Redistribution with Growth* (London: Oxford University Press, 1974), especially chapter 1 and annex. Also Roger D. Hansen, "The Emerging Challenge: Global Distribution of Income and Economic Opportunity," in *The U.S. and World Development: Agenda for Action 1975*, ed. James W. Howe (New York: Praeger, 1975), pp. 157-188.

32. See Lloyd G. Reynolds et al., "Observations on the Chinese Economy" (New Haven, Conn., 1 December 1973); also Reynolds, "China's Economy: A View from the Grass Roots," *Challenge*, vol. 17, no. 1 (March/April 1974), pp. 12-20.

33. For the opposing view—that there *is* famine in China—see Miriam and Ivan D. London, series of three articles on "The Other China" in *Worldview*, vol. 19, no. 5 (May 1976), pp. 4-11; no. 6 (June 1976), pp. 43-48; and nos. 7/8 (July/August 1976), pp. 25ff.

34. As this applies to health, see Peter S. Heller, "The Strategy of Health-Sector Planning," in *Public Health in the People's Republic of China*, ed. Myron E. Wegman (New York: J. Macy Foundation, 1973), pp. 62-107.

35. See David A. Robinson, "From Confucian Gentlemen to the New Chinese 'Political' Man," in *No Man Is Alien*, ed. J. Robert Nelson (Leiden, The Netherlands: E.J. Brill, 1971), p. 159.

36. These have been examined in several agricultural communes by Professor Neville Maxwell and his colleagues at the Institute of Commonwealth Studies, Oxford University. Much valuable information can be found in Maxwell's field notes and unpublished seminar reports to the Oxford group.

37. Simon Kuznets, "Economic Growth and Income Inequality," *The American Economic Review*, vol. 45, no. 1 (March 1955), pp. 1-28.

38. William O. Bourke, "Basic Vehicle for Southeast Asia," in *Technology and Economics in International Development* (Report of a seminar sponsored by the Agency for International Development, Washington, D.C., 23 May 1972), p. 72.

39. Quoted in Richard J. Barnet and Ronald E. Müller, *Global Reach: The Power of the Multinational Corporations* (New York: Simon and Schuster, 1974), p. 55, from an address by Gerstacker, entitled "The Structure of the Corporation," delivered at the White House Conference on *"The Industrial World Ahead,"* 7 February 1972.

40. This is the theme of Neil H. Jacoby, *Corporate Power and Social Responsibility* (New York: Macmillan, 1973).

41. P.T. Bauer, *Dissent on Development Studies and Debates in Development Economics* (Cambridge, Mass.: Harvard University Press, 1972), pp. 89-90.

42. Slogan mentioned by Professor Paul T.K. Lin in a conference on "The Current Chinese Struggle over Domestic Development Strategies" at the Annual Conference of the Canadian Society for Asian Studies, Montreal, 13 May 1975.

43. The "austerity program" has been entrusted to Jorge Cauas, disciple of Milton Friedman and the "Chicago school." See *Latin America* (London), vol. 9, no. 18 (9 May 1975), p. 139.

44. My interpretation of the Chinese approach to "austerity" (or of the "bear up, strive on" incentive) is based largely on conversations with Professor Paul T.K. Lin, whom I thank for his comments. Cf. Roger Garaudy, *Le problème chinois* (Paris: Seghers, 1967), pp. 222-227. Garaudy's private comments have also proved helpful.

45. On voluntary austerity, see Denis Goulet, *The Cruel Choice: A New Concept in the Theory of Development* (New York: Atheneum, 1971), pp. 255-263. Also Goulet, "Voluntary Austerity: The Necessary Art," *The Christian Century*, vol. 83, no. 23 (8 June 1966), pp. 748-752.

46. Jacques Ellul, "The Technological Order," in *The Technological Order*, ed. Carl F. Stover (Detroit: Wayne State University Press, 1963), p. 26.

47. For an early statement of the position, see A. Zvorikine, "The Laws of Technological Development," in *The Technological Order*, ed. Carl F. Stover, pp. 59-74. For a somewhat broader view of Soviet science, see Eugene Rabinowitch, "Soviet Science—A Survey," in *The New Technology and Human Values*, ed. John G. Burke (Belmont, Calif.: Wadsworth, 1966), pp. 360-372.

48. These include E.F. Schumacher and his associates, Robin Clarke, Peter Harper, Karl Hess, et al.

49. Alan Wells, *Picture Tube Imperialism? The Impact of U.S. Television on Latin America* (Maryknoll, N.Y.: Orbis Books, 1972), and Herbert I. Schiller, *The Mind Managers* (Boston: Beacon, 1973).

50. See Elizabeth Mann Borgese, "Report from Caracas: The Law of the Sea," *The Center Magazine*, vol. 7, no. 6 (November/December 1974), pp. 25-34.

Chapter 8 *(text pp. 167-193)*

1. François Hetman, *Society and the Assessment of Technology* (Paris: Organisation for Economic Co-operation and Development, 1973), p. 31.

2. Cf. Amílcar O. Herrera, ed., *América latina, ciencia y tecnología en el desarrollo de la sociedad* (Santiago: Editorial Universitaria, 1970); Henry Maksoud et al., *Ciência, tecnologia e desenvolvimento* (São Paulo: Editora Brasiliense, 1971); J. Leite Lopes, *Ciência e libertação* (Rio de Janeiro: Paz a Terra, 1969); and Miguel S. Wionczek, *Inversión y tecnología extranjera en América latina* (Mexico City, 1971).

3. Organization of American States, *Final Report of the Specialized Conference on the Application of Science and Technology to Latin American Development*, held in Brasília, 12-19 May 1972 (OEA/Ser. C/VI.22.1), n.d., pp. 10-11.

4. United Nations, Department of Economic and Social Affairs, *Summary of the Hearings Before the Group of Eminent Persons to Study the Impact of Multinational Corporations on Development and on International Relations* (ST/ESA/15), 1974.

5. The term is from Barbara Ward, "Development in the World Village," in *Political and Social Realities of Development: Recognition and Response*, ed. Patricia W. Blair (Dobbs Ferry, N.Y.: Oceana Publications, 1973), p. 12.

6. John D. Montgomery, *Making and Implementing Development Decisions* (Cambridge, Mass.: MIT Press, 1974), p. 11.

7. *Tapia Says: Power to the People, Proposals for a New Constitution* (Port of Spain, Trinidad: Tapia House, 1974), p. 5.

8. For a valuable summary of information problems, see W.T. Know, "Systems for Technological Information Transfer," *Science*, vol. 181, no. 4098 (3 August 1973), pp. 415-419.

9. For a summary, see Jorge Sabato and Natalio Botana, "La ciencia y la tecnología en el desarrollo de América latina," in *América latina, ciencia y tecnología en el desarrollo de la sociedad*, ed. Amílcar O. Herrera (Santiago: Editorial Universitaria, 1970), pp. 59-76. For additional works by or on Sabato, see references given at fn. 23 in chapter 4.

10. United Nations, Advisory Committee on the Application of Science and Technology to Development, *World Plan of Action for the Application of Science and Technology to Development* (ST/ECA/146), 1971, p. 51.

11. On the latter point see Pierre F. Gonod, *Clés pour le transfert technologique* (Washington, D.C.: Economic Development Institute, International Bank for Reconstruction and Development, 1974), especially pp. 353-405.

12. See *Choice and Adaptation of Technology in Developing Countries* (Paris: Development Centre, Organisation for Economic Co-operation and Development, 1974), pp. 93ff.

13. George S. Counts, *Educação para uma sociedade de homems livres na era tecnológica* (Rio de Janeiro: Centro Brasileiro de Pesquisas Educacionais, 1958), pp. 71, 79. Parallel texts in Portuguese and English.

14. See Ernesto Luis de Oliveira, Jr., *Ensino técnico e desenvolvimento* (Rio de Janeiro: Textos Brasileiros de Pedagogia, ISEB, 1959), p. 44.

15. Thomas Vietorisz, "Diversification, Linkage and Integration: Focus in the Technology Policies of Developing Countries" (Report on the proceedings of a seminar on "The Role of Small-Scale Industries in Transfer of Technology," Schloss Hernstein, Austria, 5-8 July 1973) (Paris: Development Centre, Organisation for Economic Co-operation and Development, n.d.), p. 10.

16. For Brazil, see Francisco Almeida Biato et al., *A transferência de tecnologia do Brasil* (Brasília: Instituto de Planejamento Económico e Social, 1973), pp. 215-232. For Argentina, see the bulletin of the Cámara de Diputados de la Nación, Reunión 26 (Buenos Aires), 3 May 1974, pp. 2790-2792, which describes the 1974 technology transfer law, ley n. 20.794.

17. "Twisting Whose Arm?" *The Economist*, 29 November 1975, p. 80.

18. Thomas Vietorisz, "Industrial Development Consultancy and Choice of Technology" (Paper provided at a meeting of the Organisation for Economic Co-operation and Development Working Group on Industrial Development Consultancy and Choice of Technology, Paris, 26-27 March 1973), p. 18.

19. S.A. Aluko, "Social Prerequisites for Technological Development—An African Perspective," *The Ecumenical Review* (Geneva: World Council of Churches), vol. 24, no. 3 (July 1972), p. 25.

20. For the basic texts on this legislation, see ley n. 20.794, *Ley de transferência de tecnología del exterior*, 27 September 1974, and "Mensaje del poder ejecutivo por el que se introdujo el proyecto de ley después sancionado como ley 20.794" in the bulletin of the Cámara de Diputados de la Nación, Reunión 26 (Buenos Aires), 3 May 1974, pp. 2790-2792.

21. List based on a United Nations Industrial Development Organization abstract from *Guidelines for the Acquisition of Foreign Technology in Developing Countries with Special Reference to Technology Licensing Agreements,* Background Document no. 10 (Presented to the Study Group on the Choice and Adaptation of Technology in Developing Countries, Paris, 7-9 November 1972) (Paris: Development Centre, Organisation for Economic Co-operation and Development, n.d.).

22. Eduardo Albertal and Richard L. Duncan, "Technical Cooperation Among Developing Countries: A New Approach to Multilateral Development Assistance," *International Development Review*, vol. 16, no. 4 (1974), p. 4.

23. Leon J. Rosenberg and Molly K. Hageboeck, "Management Technology and the Developing World," in *Systems Approaches to Developing Countries*, ed. M.A. Cuenod et al. (Pittsburgh: Instrument Society of America, 1973), p. 1.

24. See John W. Thomas, "Development Institutions, Projects and Aid: A Case Study of the Water Development Programme in East Pakistan," *Pakistan Economic and Social Review*, vol. 12, no. 1 (Spring 1974), pp. 77-103.

25. Written communication from John W. Thomas to author, dated 7 March 1975.

26. See Frances Stewart, "Intermediate Technology: A Definitional Discussion," in *Choice and Adaptation of Technology in Developing Countries* (Paris: Development Centre, Organisation for Economic Co-operation and Development, 1974), pp. 161-164.

27. See Government of India, Ministry of Industrial Development, Appropriate Technology Cell, *Appropriate Technology for Rapid Economic Growth* (New Delhi, 1972). Also Government of India, Ministry of Industrial Development, Development Commissioner, Small-Scale Industries, *Assistance Programme for Small-Scale Industries* (New Delhi, 1971).

28. See M. Taghi Farvar and John P. Milton, eds., *The Careless Technology: Ecology and International Development* (Garden City, N.Y.: Doubleday, 1972).

29. For examples see the *Bulletin* of the Intermediate Technology Development Group (London), no. 10 (July 1973), p. 17. Cf. Intermediate Technology Group, *Annual Report*, 1972/73. Also *Appropriate Technology for Chemical*

Industries in Developing Countries (Mexico City: National Autonomous University of Mexico, 1972).

30. See A. Banjo, "The Significance of Appropriate Technologies in Rural Development," in *Report on Development and the Dissemination of Appropriate Technologies in Rural Areas* (Berlin: German Foundation for Developing Countries, 1972), pp. 27-38.

31. For clarification of terminology, see Denis Goulet, "The Paradox of Technology Transfers," *Bulletin of the Atomic Scientists*, vol. 31, no. 6 (June 1975), pp. 39-46.

32. For review of successful applications of "soft" technologies, see *Low-Cost Technology—An Inquiry Into Outstanding Policy Issues* (Paris: Development Centre, Organisation for Economic Co-Operation and Development, 1975).

33. Source: Paul T.K. Lin's conference on "The Current Chinese Struggle over Domestic Development Strategies" at the annual conference of the Canadian Society for Asian Studies, Montreal, 13 May 1975.

34. On this see Raphie Kaplinsky, "Control and Transfer of Technology Agreements," *Institute of Development Studies Bulletin* (University of Sussex), vol. 6, no. 4 (March 1975), pp. 53-64. Also see Ashok Kapoor, *Planning for International Business Negotiation* (Cambridge, England: Ballinger, 1975).

35. See United Nations, Department of Economic and Social Affairs, *International Development Strategy* (ST/ECA/139), 1970. Italics mine.

36. For an example, see W.H. Franklin, *Code of Worldwide Business Conduct* (Peoria, Ill.: The Caterpillar Corporation, 1974).

37. One UN document speaks of rules of conduct being "codified in a multilaterally negotiated charter." See United Nations, Department of Economic and Social Affairs, *Multinational Corporations in World Development* (ST/ECA/190), 1973, p. 92.

38. See Working Group, Pugwash Conferences on Science and World Affairs, "Draft Code of Conduct on Transfer of Technology," in United Nations Conference on Trade and Development (UNCTAD), *The Possibility and Feasibility of an International Code of Conduct in the Field of Transfer of Technology* (TD/B/AC.11/L.12), 31 May 1974; UNCTAD Secretariat (Geneva), *The Possibility and Feasibility of an International Code of Conduct in the Field of Transfer of Technology* (TD/B/AC.11/22), 6 June 1974; and the Latin American official statement on technology at subgroup 3, Science and Technology Roundtable, Caracas, Venezuela, 7-11 October 1974.

39. James Greene, *The Search for Common Ground: A Survey of Efforts to Develop Codes of Behavior in International Investment*, Report no. 531 (New York: Conference Board, 1971), p. v.

40. N.T. Wang, "Towards an International Code of Conduct for Transnational Corporations" (Paper presented to the Council on Religion and

International Affairs Workshop, Aspen, Colo., 13-18 December 1974), p. 4.

41. See Denis Goulet, *The Cruel Choice: A New Concept in the Theory of Development* (New York: Atheneum, 1971), pp. 335-341.

42. David Robertson, "International Regulations for Multinational Enterprises," *Discussion Papers, International Investment and Business Studies* (University of Reading, England), no. 13 (March 1974).

43. Jack N. Behrman, "Comments on Norms for the Transfer of Technology," in *Codes of Conduct for the Transfer of Technology: A Critique* (New York: Fund for Multimanagement Education and Council of the Americas, 1974), p. 1.

44. Ibid., p. 2.

45. Constantine V. Vaitsos, "Bargaining and the Distribution of Returns in the Purchase of Technology by Developing Countries," *Institute of Development Studies Bulletin* (University of Sussex), vol. 3 (October 1970), pp. 16-23; and "Power, Knowledge and Development Policy: Relations Between Transnational Enterprises and Developing Countries" (Paper presented at the 1974 Dag Hammarskjöld Seminar, Uppsala, Sweden, 8 February 1975). In his testimony before the United Nations, Vaitsos adduces two reasons for his opposition to "packaged sales of technology." He writes: "First, many of the technological elements acquired are never understood or assimilated by the receiving country since they are undifferentiated within a package of tied components of knowledge and other inputs. They thus lead to 'pseudo-transfers' of know-how. Secondly, the acquisition of tied-in components of know-how implies that the host country will forego the development of many skills that are not proprietary or particular to the MNC and some of which could be available to or developed by the Third World." In United Nations, Department of Economic and Social Affairs, *Summary of the Hearings Before the Group of Eminent Persons to Study the Impact of Multinational Corporations on Development and on International Relations* (ST/ESA/15), 1974.

46. Government of Mexico, *Law on the Transfer of Technology and the Use and Exploitation of Patents and Trademarks* (Mexico City: n.d.), p. 4. Also see Miguel S. Wionczek, "New Mexican Legislation on Technological Transfers," *Development Digest*, vol. 11, no. 4 (October 1973), pp. 92-97.

47. On "breaking up the package," see Gonod, *Clés*, pp. 217-228, and Robert S. Merrill, "The Study of Technology," *International Encyclopedia of the Social Sciences*, 17 vols. (New York: Macmillan, 1968), vol. 15, pp. 576-589.

48. Gonod, *Clés,* p. 221.

49. Ibid., p. 226.

50. See Constantine V. Vaitsos, "The Revision of the International Patent System, View from the Third World" (Paper presented to the Secretariat of the United Nations Conference on Trade and Development, Lima, Peru,

February 1975). The basic work on patents remains the book by Edith Tilton Penrose, *The Economics of the International Patent System* (Baltimore: The Johns Hopkins Press, 1951). Cf. Penrose, "Economics and the Aspirations of Le Tiers Monde" (Inaugural lecture, School of Oriental and African Studies, University of London, 10 February 1965).

51. The International Convention for the Protection of Industrial Property, initially agreed to by eleven nations in Paris in 1883, had more than 60 signatories by the mid-1960s. The convention establishes rules and regulations governing patent and trademark rights of citizens of participatory countries. One of its most important provisions stipulates that nationals of each country are entitled to the same advantages and protective measures that other countries confer on their own citizens.

52. See chapter 2, p. 36.

53. Gonod, *Clés*, p. 200, summarizing the views of François Perroux expressed in *Pouvoir et économie* (Paris: Bordas, 1973).

54. Data for the case described were obtained in an interview held on 4 June 1975 in Washington, D.C., with Phactuel Machado-Rego, former Brazilian naval engineer who now serves as chief of the Applied Sciences Unit, Organization of American States.

55. For a general discussion see Eugene B. Skolnikoff, *The International Imperatives of Technology–The Implications of Technology for Future Development of International Organizations* (Geneva: Carnegie Endowment for International Peace, 1970).

56. The source for this, and for the following Indian examples, is Jayanta Sarkar, "India Fast Emerging as Technology Exporter," *Depthnews* (Published in Manila as "a service to members of the Press Foundation of Asia"), 7 December 1974.

57. The source for this case is an interview held on 6 April 1973 in Buenos Aires with Lt. Colonel Julio Delucchi, chief of the Department of Mines and Geology, Fabricaciones Militares.

58. Constantine V. Vaitsos, "Summary of the Written and Oral Statement," in United Nations, Department of Economic and Social Affairs, *Summary of the Hearings Before the Group of Eminent Persons to Study the Impact of Multinational Corporations on Development and on International Relations* (ST/ESA/15), 1974, p. 398.

59. Robert N. Seidel, *Toward an Andean Common Market for Science and Technology* (Ithaca, N.Y.: Cornell University, Program on Policies for Science and Technology in Developing Nations, 1974), p. ix.

60. Raúl Prebisch, *Change and Development, Latin America's Great Task* (Washington, D.C.: Inter-American Development Bank, 1970), p. 5.

61. The complete estimate is 70% in the United States, 28% in other developed market economies, and 2% in developing countries, according to United Nations, Department of Economic and Social Affairs, *Science and Technology for Development: Proposals for the Second United Nations Development Decade* (ST/ECA/133), 1970, p. 23.

62. *Meeting the Challenge of Industrialization: A Feasibility Study for an*

International Industrialization Institute (Washington, D.C.: National Academy of Sciences, 1973).

63. Walter A. Chudson, *The International Transfer of Commercial Technology to Developing Countries*, Research Reports, no. 13 (New York: United Nations Institute for Training and Research, 1971), p. 53.

64. Seidel, *Toward an Andean Common Market,* pp. 10-14. Cf. *Andean Pact: Definition, Design, and Analysis* (New York: Council of the Americas, 1973), passim.

65. See Thomas Allen, "The International Technological Gatekeeper," *Technology Review*, vol. 73, no. 5 (March 1971), pp. 3-9.

66. Jan Tinbergen, "Report of General Meeting on RIO [Reviewing the International Order] Project" (11-13 December 1974), Club of Rome Document RIO-17, p. 9.

67. Orville L. Freeman, "Multinational Companies and Developing Countries: A Social Contract Approach," *International Development Review*, vol. 16, no. 4 (1974), pp. 17-19. Cf. André van Dam, "A Hearing Aid, or How to Amplify the Dialogue Between Multinational Corporations and the Developing Countries" (Paper delivered at the annual conference of the Society for International Development, Abidjan, Ivory Coast, August 1974).

68. In addition to the two UNCTAD documents cited in fn. 38, this chapter, see also the complete works cited in fns. 40 and 43.

69. See Jorge A. Sabato and Oscar Wortman, "Informe preliminar: apertura del paquete tecnológica para la central núclear de Atucha (Argentina)," in *Methods of Evaluation of Technology* (Washington, D.C.: Organization of American States, 1974), pp. 29-95; Sabato, "Atomic Energy in Argentina: A Case History," *World Development*, vol. 1, no. 8 (August 1973), pp. 23-38.

70. Some authors have even suggested broader international cooperation. See Ithiel de Sola Pool, "A Proposal for a New International Technology Transfer Institute," *World Future Society Bulletin* (May-June 1975), pp. 5-7.

71. Francisco R. Sagasti and Mauricio Guerrero, *El desarrollo científico y tecnológico de América latina* (Buenos Aires: BID/INTAL, 1974), p. 98.

72. On this see the interesting remarks of Gustav Ranis, "Some Observations on the Economic Framework for Optimum LDC Utilization of Technology," in *Technology and Economics in International Development* (Report of a seminar sponsored by the Agency for International Development, Washington, D.C., 23 May 1972), pp. 34ff.

73. Suitable agricultural research must not, however, be neglected. On this see Montague Yudelman, Gavan Butler, and Ranadev Banerji, *Technological Change in Agriculture and Employment in Developing Countries* (Paris: Development Centre, Organisation for Economic Co-operation and Development, 1971), pp. 143ff.

74. Roy A. Matthews, "The International Economy and the Nation State," *Columbia Journal of World Business*, vol. 6, no. 6 (November-December 1971), pp. 51-60.

75. World Intellectual Property Organization (Geneva), *WIPO Publication No. 400 (E)*, 1972/73, p. 8.

76. P.C. Trussell, "Some Reflections," in *Choice and Adaptation of Technology in Developing Countries: An Overview of Major Policy Issues* (Paris: Development Centre, Organisation for Economic Co-operation and Development, 1974), p. 158. Trussell is secretary general of the World Association of Industrial and Technological Research Organizations, a newly created inter-governmental institution aiming at the building of a network of communication among industrial and technological research institutions in developing countries.

77. "The top executives of most of the multinationals are talking a language which is quite revolutionary and which would have been inconceivable ten years ago—quality of life, environment, social responsibility, services, leisure, enrichment, participation, job satisfaction," according to José de Cubas, senior vice president of Westinghouse, as quoted in Richard J. Barnet and Ronald E. Müller, *Global Reach: The Power of the Multinational Corporations* (New York: Simon and Schuster, 1974), p. 65.

78. Raymond Vernon, "Multinational Enterprises in Developing Countries: An Analysis of National Goals and National Policies" (Draft prepared for United Nations Industrial Development Organization, 8 March 1974), p. 9.

79. The group was created in accord with the Twenty-Seventh UN General Assembly Resolution 2974 (14 December 1972) jointly with the Governing Council of the United Nations Development Programme (UNDP). See UNDP, *Report of the First Session of the Working Group on Technical Cooperation among Developing Countries* (DP/WGTC/IV), 25 July 1973.

80. See UNDP Documents DP/WGTC/II (14 June 1973); SC/TECH./AC.1/2 (20 September 1973); DP/WGTC/L.2 (4 December 1973); DP/WGTC/L.3 (22 January 1974); DP/WGTC/L.4 (20 March 1974); DP/WGTC/L.6 (29 March 1974); DP/69 (20 May 1974).

81. For an example of the latter, see United Nations Industrial Development Organization (UNIDO), Textile Quality Control Centre in Alexandria (Project EGY-69-562), which conducts chemical testing and standard laboratory tests for fibres and textile products, recommends further laboratory tests, and trains Egyptian technicians.

82. For a brief statement of the nature and objectives of the project, see Organization of American States, *Informe final: proyecto piloto de transferencia de tecnología* (SG/P.1, PPTT/34), 1975.

83. The "competitive technology" marketplace is organizing its commercial exhibits ever more systematically and in more centralized fashion. For one recent illustration of a huge fair for technology sellers and buyers, see the special issue of *UNIT*, December 1975, which discusses the Technology II 1976 World Fair [to be] held in Chicago on 26 February 1976. *UNIT* is published in Ormond Beach, Florida.

84. André van Dam, "The Triumph of the Horse," *Worldview*, vol. 18, no. 12 (December 1975), p. 30.

85. Ibid.

86. For an interesting view on how the United Nations should change to cope "systemically" with the world, see Harlan Cleveland, "The U.S. vs. the U.N.?" *The New York Times*, 4 May 1975.

Chapter 9 *(text pp. 195-232)*

1. C. Fred Bergsten, ed., *The Future of the International Economic Order: An Agenda for Research* (Lexington, Mass.: D.C. Heath, 1973); Jack N. Behrman, *Toward a New International Economic Order* (Paris: The Atlantic Institute for International Affairs, 1974); and Jagdish N. Bhagwati, *Economics and World Order from the 1970s to the 1990s* (New York: The Free Press, 1972).

2. See Denis Goulet, "Development and the International Economic Order," *International Development Review*, vol. 16, no. 2 (1974), pp. 10-16.

3. Robert O. Keohane and Joseph S. Nye, Jr., eds., *Transnational Relations and World Politics* (Cambridge, Mass.: Harvard University Press, 1973). Cf. Richard A. Falk, "Another Look at 'Development and the International Economic Order,' " *International Development Review*, vol. 16, no. 4 (1974), pp. 19-20.

4. See Tibor Mende, *From Aid to Re-Colonization: Lessons of a Failure* (New York: Pantheon, 1973); Judith Hart, *Aid and Liberation: A Socialist Study of Aid Policies* (London: Gallancz, 1973); John White, *The Politics of Aid* (New York: St. Martin's Press, 1974); and Denis Goulet and Michael Hudson, *The Myth of Aid: The Hidden Agenda of the Development Reports* (New York: IDOC/North America, 1971).

5. Gunnar Myrdal, *The Challenge of World Poverty* (New York: Pantheon, 1970), p. 277. Italics are Myrdal's.

6. Cf. Irving Louis Horowitz, "A More Equal World," *Worldview*, vol. 18, no. 12 (December 1975), p. 17: "...revolution within the structure of capitalism....They do not involve a fundamental change in socioeconomic structures at the international level."

7. Mende, *From Aid to Re-Colonization*, pp. 87-129.

8. The image of "upstream" and "downstream" is a familiar one. Obviously, if a polluter empties waste up-river or upstream, all those located farther downstream suffer from the contaminating action. Similarly, if intensive fishing is carried on upstream, there is little left to catch for those positioned below them. The same applies to access to resources, whether mineral ores or commodities: if certain institutions and interests have untrammeled extraction, gathering, or processing rights, the distributional mechanism is vitiated at its source. Consequently, to plead for access by poor to resources upstream means giving institutional legitimacy and reality to their claims on resources before these are appropriated by corporations, governments, banks, or other entities.

9. See Seyom Brown, *New Forces in World Politics* (Washington, D.C.: The Brookings Institution, 1974), particularly chapter 10.

10. See Richard A. Falk, "What's Wrong with Henry Kissinger's Foreign Policy?" Policy Memorandum no. 39 (Princeton, N.J.: Princeton University, Center of International Studies, July 1974).

11. Richard A. Falk, "On Transnational Institutions," *Center Report*, vol. 7, no. 2 (April 1974), p. 28.

12. See Soedjatmoko, "Nationalism and Internationalism: A Third World View," *Worldview*, vol. 18, no. 6 (June 1975), pp. 28-33.

13. Fouad Ajami, "The Global Populists: Third-World Nations and World-Order Crises," Research Monograph no. 41 (Princeton, N.J.: Princeton University, Center of International Studies, May 1974), p. 15.

14. Paul J. Braisted, Soedjatmoko, and Kenneth W. Thompson, eds., *Reconstituting the Human Community* (New Haven, Conn.: The Hazen Foundation, 1972), p. 11.

15. See Teresa Hayter, *Aid as Imperialism* (Middlesex, England/Baltimore: Penguin Books, 1971).

16. The "Santiago Statement" declares, *inter alia*, that "the Forum is a completely independent organization, with no institutional affiliations, and is open to all the leading social scientists from the Third World with a predominant interest in the development of their societies....The Forum offers its hand and its support to all liberal and progressive elements all over the world in working towards the establishment of a just and responsive world order." (Meeting held in Santiago, Chile, 23-25 April 1973.)

At a subsequent meeting in Karachi, Pakistan, 5-10 January 1975, the Forum issued its constitution and a formal "Communiqué of the Third World Forum." See "The Third World Forum: Intellectual Self-Reliance," *International Development Review*, vol. 17, no. 1 (1975), pp. 8-13.

17. On development language see Denis Goulet, "Development...or Liberation?" *International Development Review*, vol. 13, no. 3 (1971), pp. 6-10.

18. For a useful overview see David H. Blake, ed., *The Multinational Corporation*, special issue of *The Annals of the American Academy of Political Science*, vol. 403 (September 1972).

19. Orville L. Freeman, "Remarks" (Delivered at a conference on "New Structure for Economic Interdependence," United Nations, New York, 15 May 1974), mimeographed, p. 4.

20. A. Barber, "Emerging New Power: The World Corporation," *War/Peace Report*, October 1968, p. 7, cited by Oswaldo Sunkel in "Big Business and 'Dependencia,' " *Foreign Affairs*, vol. 50, no. 3 (April 1972), pp. 52-53.

21. Jack N. Behrman speaks of a "United Nations statute of incorporation." See his *Conflicting Restraints on the Multinational Enterprise* (New York: Council of the Americas/Fund for Multinational Management Education, 1974), p. 61.

22. See Arnold Toynbee, Orville Freeman, Aurelio Peccei, and Eldridge Haines, "Will Businessmen Unite the World?" *Center Report*, vol. 4, no. 2 (April 1971), pp. 8-10.

23. Franklin Tugwell explores the possibility in "The 'Soft States' Can't Make It," *Center Report*, vol. 5, no. 5 (December 1972), pp. 15-17. A more complete statement is found in Tugwell, "Modernization, Political Development, and the Study of the Future," in *Political Science and the Study of the Future*, ed. Albert Somit (New York: Dryden, 1974), pp. 155-173.

24. Freeman ("Remarks," p. 3) asserts that the international corporation is such an effective instrument of economic growth because it "does not merely transfer resources—which it also does—but custom crafts this transfer into a package of productive factors. Capital, technology and management skills are hammered together by the international corporation to meet the needs of a given project, whatever that project may be."

25. Statements based on conversations with Professor Raymond Vernon during the period 1973-74. Cf. Raymond Vernon, "Conflict and Resolution Between Foreign Direct Investors and Less-Developed Countries," *Public Policy*, vol. 17 (1968), pp. 333-351.

26. On the International Federation of Institutes for Advanced Studies, see "IFIAS: The Top People's Problem Posers," *Nature*, vol. 256 (10 July 1975), pp. 80-82.

27. Diana Crane, "Transnational Networks in Basic Science," in *Transnational Relations and World Politics,* ed. Robert O. Keohane and Joseph S. Nye, Jr. (Cambridge, Mass.: Harvard University Press, 1973), p. 235.

28. As brilliantly depicted in Arthur Koestler's recent novel *The Call Girls* (New York: Random House, 1973).

29. For a searching presentation of new possibilities, see Johan Galtung, "Nonterritorial Actors and the Problem of Peace," in *On the Creation of a Just World Order*, ed. Saul H. Mendlovitz (New York: The Free Press, 1975), pp. 151-188.

30. Ibid., p. 176.

31. On this see Richard A. Falk, *A Study of Future Worlds* (New York: The Free Press, 1975).

32. For details see Denis Goulet, "The Political Economy of the Image Industries," *IDOC International (North American Edition)*, 13 November 1971, pp. 38-62. Also see Herbert I. Schiller, "Communications Satellites: A New Institutional Setting," *Bulletin of the Atomic Scientists*, vol. 23, no. 4 (April 1967), pp. 4-8.

33. Gordon L. Weil, ed., *Communicating by Satellite: An International Discussion* (New York: Carnegie Endowment for International Peace/Twentieth Century Fund, 1969), p. 1. Cf. Herbert I. Schiller, "The Sovereign State of COMSAT," *The Nation*, 25 January 1965, pp. 71-76.

34. Weil, *Communicating by Satellite*, p. 3.

35. COMSAT [Communications Satellite Corporation] *Annual Report*, 1974, p. 4.

36. COMSAT 1974 report to the President and the Congress, *New Ventures for a New Decade* (Washington, D.C.: COMSAT, 1974), p. ii.

37. For detailed treatment see Richard A. Falk, "A New Paradigm for International Legal Studies: Prospects and Proposals," *The Yale Law Journal*, vol. 84, no. 5 (April 1975), pp. 969-1021.

38. One highly provocative exception is the work of Everett Reimer, "Alternative Futures for the World," unpublished manuscript dated June 1973.

39. Diana Crane, "Transnational Networks," p. 237.

40. *The World's Earth Stations for International Satellite Communications* (Washington, D.C.: COMSAT, 1974), p. 2.

41. Charles R. Dechert, "A Pluralistic World Order," in *Proceedings of the American Catholic Philosophical Association* (Washington, D.C., 1973), p. 167.

42. Mihajlo Mesarovic and Eduard Pestel, *Mankind at the Turning Point: The Second Report to the Club of Rome* (New York: Dutton/Reader's Digest Press, 1974), p. 34. Cf. Louis René Bores and Harry R. Targ, *Reordering the Planet: Constructing Alternative World Futures* (Boston: Allyn and Bacon, 1974), and Ervin Laszlo, *A Strategy for the Future: The Systems Approach to the World Order* (New York: Braziller, 1974.)

43. Lincoln P. Bloomfield, "Is the UN Relevant to the 1970s?" *VISTA* (Magazine of the United Nations Association of the USA), vol. 5, no. 6 (May/June 1970), pp. 106-111.

44. Speech by Henry Kissinger, Sixth Special Session of the United Nations General Assembly, New York, 15 April 1974.

45. Neil H. Jacoby, *Corporate Power and Social Responsibility* (New York: Macmillan, 1973), pp. 121-122.

46. Houari Boumédiene, *Message to the Secretary-General of the United Nations*, dated 30 January 1974.

47. Ali A. Mazrui, "The New Interdependence: From Hierarchy to Symmetry," in *The U.S. and World Development: Agenda for Action 1975*, ed. James W. Howe (New York: Praeger, 1975), p. 118.

48. Ibid., p. 20.

49. Falk, *A Study of Future Worlds*, pp. 174-223.

50. The work of Falk and his colleagues in the World Order Models Project (WOMP) is predicated on the need to harness the energies of thinkers, activists, and political leaders to promote these four values in their efforts to affect the evolutionary outcome of world order: the minimization of large-scale violence, the maximization of social and economic well-being, the realization of fundamental human rights and the conditions of political justice, and the maintenance and rehabilitation of ecological quality. On this see Falk, *A Study of Future Worlds*, pp. 11-34.

Similar values are advocated by numerous others interested in futures studies, ecological integrity, safeguarding the carrying capacity of the planet, and technology assessment.

51. Unless, of course, the "triage" policy is adopted by big powers. See James W. Howe and John W. Sewell, "Let's Sink the Lifeboat Ethics," *Worldview*, vol. 18, no. 10 (October 1975), pp. 13-18.

52. Falk, "On Transnational Institutions," p. 28.

53. C. Fred Bergsten, "The Threat from the Third World," *Foreign Policy*, no. 11 (Summer 1973), pp. 102-104. Cf. Bergsten, "The Threat Is Real," *Foreign Policy*, no. 14 (Spring 1974), pp. 84-90; and Bergsten, "The Response to the Third World," *Foreign Policy*, no. 17 (Winter 1974-75), pp. 3-34.

54. Cf. Daniel P. Moynihan, "The United States in Opposition," *Commentary*, vol. 59 (March 1975), pp. 31-44.

55. Zbigniew Brzezinski, *Between Two Ages: America's Role in the Technetronic Era* (New York: Viking, 1970), p. 160.

56. Cf. Erich Fromm, *Escape from Freedom* (New York: Holt, Rinehart & Winston, 1963).

57. Jack N. Behrman, *Toward a New International Economic Order* (Paris: The Atlantic Institute for International Affairs, 1974), p. 10.

58. Henry C. and Mabel I. Wallich, "Economics and Ideology: A Three-Dimensional View," *Economic Impact*, no. 8 (Washington, D.C.: US Information Agency, 1974), p. 59.

59. The phrase, from Paul Rosenstein-Rodan, is quoted in Richard J. Barnet and Ronald E. Müller, *Global Reach: The Power of the Multinational Corporations* (New York: Simon and Schuster, 1974), p. 188.

60. Barnet and Müller, *Global Reach*, p. 13.

61. David Ignatious, "Taming the Beast: The Multinationals," *The New Republic*, vol. 171, no. 11 (14 September 1974), p. 19.

62. A.A. Fatouros, "The Computer and the Mud Hut: Notes on Multinational Enterprise in Developing Countries," *Columbia Journal of Transnational Law*, vol. 10, no. 2 (Fall 1971), p. 341.

63. Cited in Barnet and Müller, *Global Reach*, p. 16.

64. For example, see André van Dam, "The Multinational Corporation vis-à-vis Societies in Transformation: The Case for Intermediate Technology in Developing Countries," *Technological Forecasting and Social Change*, vol. 5 (1973), pp. 281-293; van Dam, "Global Development: From Confrontation to Cooperation?" *Planning Review*, vol. 2, no. 5 (August/September 1974), pp. 1-4, 20-23; van Dam, "The Triumph of the Horse," *Worldview*, vol. 18, no. 12 (December 1975), pp. 27-30.

65. See Ghita Ionescu and Ernest Gellner, eds., *Populism, Its Meanings and National Characteristics* (London: Weidenfeld and Nicolson, 1969).

66. Paul G. Hoffman, comment on jacket of Denis Goulet's *The Cruel*

Choice: A New Concept in the Theory of Development (New York: Atheneum, 1971).

67. Fouad Ajami, "The Global Populists: Third-World Nations and World-Order Crises," Research Monograph no. 41 (Princeton, N.J.: Princeton University, Center of International Studies, May 1974), p. 1.

68. Falk, "On Transnational Institutions," p. 28.

69. The philosophical implications of "massifications" are discussed in Gabriel Marcel, *Man Against Mass Society* (Chicago: Henry Regnery, 1962); and Jose Ortega y Gasset, *The Revolt of the Masses* (New York: Norton, 1932).

70. Falk, "A New Paradigm," pp. 969-1021.

71. Expounded at length in Falk, *A Study of Future Worlds,* pp. 277-349.

72. The theory of the revolutionary "focus" is propounded by Regis Debray, *Revolution in the Revolution?* trans. Bobbye Ortiz (New York: Grove Press, 1967).

73. Both illusions are luminously analyzed in Barrington Moore, Jr., *Reflections on the Causes of Human Misery and upon Certain Proposals to Eliminate Them* (Boston: Beacon, 1972), pp. 13ff.

74. "Bribes and Arms Sales," editorial, *The Washington Post*, 13 June 1975.

75. "Nuclear Madness," editorial, *The New York Times*, 13 June 1975. Italics mine. For greater detail on the Brazil/West Germany nuclear arrangements, see *Latin America Economic Report*, vol. 3, no. 24 (20 June 1975), p. 93.

76. Albert Camus, *The Rebel* (New York: Vintage, 1956), p. 304.

77. On nonviolent politics see Joan Valerie Bondurant, *Conquest of Violence: The Gandhian Philosophy of Conflict* (Berkeley: University of California Press, 1965), and Gene Sharp, *The Politics of Nonviolent Action* (Boston: Porter Sargent, 1973).

78. Richard A. Falk, "A New Paradigm," pp. 973-974.

79. Edward Banfield, *The Moral Basis of a Backward Society* (Glencoe, Ill.: The Free Press, 1965).

80. François Perroux, *La coéxistence pacifique* (Paris: Presses universitaires de France, 1958), vol. 3, p. 623. Translation mine.

81. As warrant for this view, one notes the proliferation of such associations of "world-minded" people as Planetary Citizens, the World Federalists, Amnesty International, et al.

82. The attitudes, values, and aspirations of corporate managers are studied in Peter Cohan, *The Gospel According to the Harvard Business School* (Garden City, N.Y.: Doubleday, 1974).

83. John P. Marquand, *Sincerely, Willis Wayde* (Boston: Little, Brown, 1955).

84. This conclusion is confirmed by Michael Maccoby (in an interview conducted in Washington, D.C., 25 November 1974), who states that a good corporate manager must combine four qualities: technical training as an innovator, the ability to cooperate on a team, a fierce competitive drive, and the ability to be competitive without being destructive.

85. Rabindranath Tagore, cited in Louis-Joseph Lebret, *Développement-révolution solidaire* (Paris: Les Editions Ouvrières, 1967), p. 52.

86. Antonio Martins Filho, *O universal pelo regional, definição de uma política universitaria* (Fortaleza, Brazil: Imprensa Universitaria do Ceará, 1965).

87. Karl Mannheim, *Freedom, Power and Democratic Planning* (London: Routledge & Kegan Paul, 1951), p. 62.

88. On the difference between *leadership* and *rulership*, see Paul T.K. Lin, "Development Guided by Values: Comments on China's Road and Its Implications," in *On the Creation of a Just World Order*, ed. Saul H. Mendlovitz (New York: The Free Press, 1975), pp. 268ff.

89. For a typical example see Albert Waterston, "A Viable Model for Rural Development," *Finance and Development*, vol. 2, no. 4 (December 1974), pp. 22-25.

90. For an initiation to Kaplan's thought, see his "The Power Structure in International Relations," *International Social Science Journal*, vol. 26, no. 1 (1974), pp. 95-108; "Hacia un modelo mundial alternativo: lineamientos sociopolíticos," *Comercio Exterior* (Mexico), July 1973, pp. 689-703; "Hacia una alternative socialista para la crisis argentina," *Voz libre* (Buenos Aires), January/February 1973, pp. 5-8; "La ciudad latinoamericana como factor de transmisión de poder socioeconómico y político hacia el exterior durante el periodo contemporaneo" (Reprint from the proceedings of the Thirty-Ninth International Congress of "Americanistas," Lima, Peru, 1970), pp. 219-256; *La investigación latinoamericana en ciencias sociales* (Mexico City: El Colegio de Mexico, 1973). I thank Professor Kaplan for illuminating conversations held in Lima, Santiago, and Buenos Aires on his views.

91. See Severyn T. Bruyn, Norman J. Faramelli, and Dennis A. Yates, *An Ethical Primer on the Multinational Corporation* (New York: IDOC/North America, 1973). Also *Corporate Action Guide* (Washington, D.C.: Interaction Center, 1974).

92. W. Michael Blumenthal, "MNCs Under Fire: The Challenge to Chief Executive Officers," in Business International Corporation's *Weekly Report to Managers of Worldwide Organizations*, 6 June 1975, p. 181.

93. Ibid.

94. Ibid.

95. Ibid.

96. Adolf A. Berle, Jr., *The 20th Century Capitalist Revolution* (New York: Harcourt, Brace & World, 1954), p. 166.

97. This phenomenon clearly explains the incipient vogue of "social auditing

for corporations." On this see Raymond A. Bauer and Dan H. Fenn, Jr., "What Is a Corporate Social Audit?" *Harvard Business Review* (January-February 1973), pp. 37-48. Also, Peter Merrill, "Social Accounting for Decision-Making: A Survey of Approaches to Date" (Paper presented at The Social Measurement Workshop, Boston, 18-20 November 1973).

98. W.T. Stace, "Man Against Darkness," in *The New Technology and Human Values*, ed. John C. Burke (Belmont, Calif.: Wadsworth, 1966), p. 51.

99. The phrase is from William Kuhns, *The Post-Industrial Prophets: Interpretations of Technology* (New York: Harper & Row, 1971), p. 249.

100. For interesting remarks on this "liberal" view of participation, see Robert A. Packenham, *Liberal America and the Third World* (Princeton, N.J.: Princeton University Press, 1973), pp. 146-160.

101. Jan B. Luytjes, "Road to Lunacy" (Working Document no. 1 of the World Resource Management Conference, Florida International University, November 1974), mimeographed, p. 1.

102. Ibid.

103. Eugene B. Skolnikoff, "The Governability of Complexity," in *Growth in America,* ed. Chester L. Cooper (Westport, Conn.: Greenwood Press, 1976). Quotation in text is found on p. 10 of pre-publication mimeographed manuscript.

104. For a review of the operations of markets in pre-industrial societies, see George Dalton, ed., *Tribal and Peasant Economies: Readings in Economic Anthropology* (Garden City, N.Y.: Doubleday, 1967).

105. Mannheim, *Freedom, Power and Democratic Planning*, p. 191.

106. The conventional example invoked to disprove this assertion is Sweden. Yet there, too, the market wreaks its own special brand of havoc, not least of which is mystification of the political process. For a penetrating critique, see R.H. Weber, "Sweden Inc.: The Total Institution," *Worldview*, vol. 18, no. 3 (March 1975), pp. 20-30.

107. See, for instance, Richard A. Jackson, ed., *The Multinational Corporation and Social Policy* (New York: Praeger, 1974).

108. John Kenneth Galbraith, *Economic Development in Perspective* (Cambridge, Mass.: Harvard University Press, 1962), p. 43. Italics are Galbraith's.

109. Thomas Aitken, *The Multinational Man: The Role of the Manager Abroad* (New York: John Wiley & Sons, 1973), p. 135.

110. Soedjatmoko, "Nationalism and Internationalism," p. 32.

111. André Malraux, *Malraux par lui-même* (Paris: Editions du Seuil, 1953), p. 179. Translation mine.

Conclusion *(text pp. 235-251)*

1. Margaret Mead, "Establishing a Shared Culture," *Daedalus*, vol. 94, no. 1 (Winter 1965), p. 143.

2. In "The Place of Intercultural Relations in the Study of International Relations" (Paper presented at the International Studies Association meeting, St. Louis, Missouri, 22 March 1974), A. Roy Preiswerk defines culture (p. 2) as "a totality of values, institutions and forms of behavior transmitted within a society, as well as the material goods produced by man. It must be noticed that this wide concept of culture covers *Weltanschauung*, ideologies and cognitive behavior." Cf. E. Sapir, "Culture, Genuine and Spurious," *The American Journal of Sociology*, vol. 29 (January 1924), pp. 401-429.

3. Jacques Ellul, "The Technological Society: A Dialogue," *The Nation*, 24 May 1965, p. 567.

4. Ibid., p. 568.

5. Jacques Ellul, *Hope in a Time of Abandonment* (New York: The Seabury Press, 1973), p. 232.

6. Bertrand de Jouvenel, "Musings on Technology" (Working paper presented at the Center for the Study of Democratic Institutions, Santa Barbara, Calif., 5 January 1966), mimeographed, p. 1. Cf. Jouvenel, *The Art of Conjecture* (New York: Basic Books, 1967).

7. Everett Reimer, "Alternative Futures for the World," unpublished book manuscript (n.p., June 1973), p. 125.

8. Jacques Ellul, "Ideas of Technology," in *The Technological Order*, ed. Carl F. Stover (Detroit: Wayne State University Press, 1963), pp. 26-27.

9. On ethics as a "means of the means" see Denis Goulet, "Ethical Strategies in the Struggle for World Development," in *Global Justice and Development: Report of the Aspen Interreligious Consultation* (Washington, D.C.: Overseas Development Council, 1975), pp. 42ff.

10. See Charles M. Savage, *Work and Meaning: A Phenomenological Inquiry* (Springfield, Va.: US Department of Commerce, National Technical Information Service, 1973). [cf. fn. 41 on p. 275]

11. *Technology Assessment for the Congress*, staff study of the Subcommittee on Computer Services of the Committee on Rules and Administration, United States Senate (Washington, D.C.: US Government Printing Office, 1972), p. 1.

12. Alexander King, "Foreword" to François Hetman, *Society and the Assessment of Technology* (Paris: Organisation for Economic Co-operation and Development, 1973), p. 7.

For a bibliographical introduction to technology assessment, consult the study cited in the previous note, pp. 83-105. Also useful are the bibliographies contained in John D. Montgomery, *Technology and Civic Life: Making and Implementing Development Decisions* (Cambridge, Mass.: MIT Press, 1974), p. 193; National Academy of Engineering, Committee on Public Engineering Policy, *A Study of Technology Assessment* (Report to the Committee on Science and Astronautics, US House of Representatives) (Washington, D.C.: US Government Printing Office, 1969); Raphael G. Kasper, ed., *Technology Assessment: Understanding the Social Consequences of Technological Applications* (New York: Praeger, 1973), especially articles by Mayo and Huddle; Vary T. Coates, *Technology and Public Policy* (Washington, D.C.: George

Washington University, Program of Policy Studies in Science and Technology, 1972); and Martin V. Jones, ed., *A Technology Assessment Methodology*, 6 vols. (Falls Church, Va.: Mitre Corporation, 1973), vol. 1, p. 22, in turn based on National Academy of Engineering, *A Study of Technology Assessment*, p. 25.

13. Remarks by Helmut Krauch in Anthony De Reuck, Maurice Goldsmith, and Julie Knight, eds., *Decision Making in National Science Policy* (Boston: Little, Brown, 1968), p. 96. Cf. C. West Churchman, B. Kohler, and H. Krauch, *Experiments on Inquiring Systems* (Berkeley: University of California Press, 1967).

14. Sherry R. Arnstein and Alexander N. Christakis, "Perspectives on Technology Assessment" (Draft report of a workshop sponsored by the Academy for Contemporary Problems and the National Science Foundation, 28-30 July 1974), chapter 16, p. 1. The same authors present their views more fully in *Perspectives on Technology Assessment* (Jerusalem: Science and Technology Publishers, 1975).

15. Helmut Krauch in De Reuck et al., eds., *Decision Making*, p. 97.

16. See Helmut Krauch, "Public Control of Government Planning" (Paper discussed at Center for the Study of Democratic Institutions, Santa Barbara, Calif., 19-20 March 1969). Also Helmut Krauch and C. West Churchman, "Experiments, Experience and Planning" (Paper discussed at same center, 21 March 1969).

17. Valentina Borremans and Ivan Illich, "La necesidad de un techo común: el control social de la tecnología" (Paper prepared for a seminar at the Centro Intercultural de Documentación, Cuernavaca, Mexico, January 1972), p. 1.

18. François Hetman, *Society and the Assessment of Technology*, p. 377.

19. On *praxis* see Richard J. Bernstein, *Praxis and Action* (Philadelphia: University of Pennsylvania Press, 1971).

20. Arthur Harkins, "Toward a Technology of Values" (Paper delivered at the Rome Special World Conference on Futures Research, September 1973), mimeographed, p. 1. Harkins refers to Magoroh Maruyama, "A New Logical Model for Future Research" (Paper presented at the Third World Research Conference, Bucharest, Romania, 3-10 September 1972), and to Maruyama, "Symbiotization of Cultural Heterogeneity: Scientific, Epistemological and Esthetic Bases," *1972 American Anthropological Association Cultural Futuristic Pre-Conference Volume* (Minneapolis: University of Minnesota, 1972).

21. Harkins, "Toward a Technology of Values," p. 15.

22. Ian I. Mitroff and Murray Turoff, "The Whys Behind the Hows," *Institute of Electrical and Electronics Engineers Spectrum*, vol. 10, no. 3 (March 1973), p. 68.

23. Ibid., p. 71.

24. Ibid.

25. Christopher Dawson, *Dynamics of World History* (New York: Sheed and Ward, 1957), p. 364.

26. Not the least insidious of modern idolatries is the worship of techniques of spiritual emancipation. See Chogyam Trungpa, *Cutting Through Spiritual Materialism* (Berkeley, Calif.: Shambhala Publications, 1973).

27. Cf. Denis Goulet, "Needed: A Cultural Revolution in the U.S.," *The Christian Century*, vol. 91, no. 30 (4 September 1974), pp. 816-818.

28. William Kuhns, *The Post-Industrial Prophets: Interpretations of Technology* (New York: Harper & Row, 1971), p. 259.

29. Ibid., p. 260.

30. Pierre Ducassé, *Les techniques et le philosophe* (Paris: Presses universitaires de France, 1958), p. 7-20.

31. Title of an important book by Kierkegaard. See *The Sickness Unto Death*, trans. Walter Lowrie (Princeton, N.J.: Princeton University Press, 1941).

32. Søren Kierkegaard, *The Present Age* (New York: Harper & Row, 1962), pp. 39-40.

33. James Harvey Robinson, "Civilization and Culture," in *Encyclopedia Britannica*, 14th ed., 23 vols. (Chicago: William Benton, 1969), vol. 5, p. 831.

34. Arnold J. Toynbee, *Civilization on Trial* (New York: Oxford University Press, 1948), p. 158.

35. John White, "What Is Development? And for Whom?" (Paper presented at Quaker Conference on "Motive Forces in Development," Hammamet, Tunisia, April 1972), p. 41.

36. Toynbee, *Civilization on Trial*, p. 86.

37. Ibid., p. 91.

38. Cf. Toynbee, *Civilization on Trial*, pp. 70, 158.

39. Lord Ritchie-Calder, *After the Seventh Day: The World Man Created* (New York: Simon and Schuster, 1961), p. 19. For a more detailed and scientific portrait of the adaptability of the Innuits, cf. Sixten S.R. Haraldson, *Evaluation of Alaska Native Health Service* (Report of a study trip, December 1972/January 1973, prepared for the World Health Organization, Geneva).

40. See, for example, the address of Henry Kissinger before the St. Louis World Affairs Council, 12 May 1975. (Text available from US Department of State, Bureau of Media Affairs, Washington, D.C.)

Although widely hailed as being "generous" and attentive to Third World needs, the speech nonetheless betrays the assumption that 25% of the world's population has every right to 75% of its resources and that the United States, in particular, need feel no uneasiness in controlling at least 30% of these resources to satisfy its mere 6% of the world's peoples. No mention is made of the risk that the pursuit by Americans of still further material wealth can endanger the biosphere and inflict physical and genetic damage on future generations.

41. Raymond Aron, *Dix-huit leçons sur la société industrielle* (Paris: Gallimard, 1962), p. 361. Translation mine.

42. Jacques Ellul, *The Technological Society*, p. 424.

43. Ibid., p. 427.

44. Scott Buchanan, "Technology as a System of Exploitation," in *The Technological Order*, ed. Carl F. Stover (Detroit: Wayne State University Press, 1963), p. 159.

45. A.B. Zahlan, "Cultural Change and Cultural Transfer: A Preliminary Assessment of Present Conditions" (n.p., 6 September 1973), mimeographed, p. 2.

46. The phrase is Lord Ritchie-Calder's, in his *After the Seventh Day*, p. 427.

47. Quoted in Nicholas Wade, "E.F. Schumacher: Cutting Technology Down to Size," *Science*, vol. 189, no. 4198 (18 July 1975), p. 199.

48. Arnold J. Toynbee, *A Study of History*, 10 vols., abridgement by D.C. Somervell, in 2 vols. (New York: Dell, 1965), vol. 1, pp. 59, 379, 382.

49. Ibid., vol. 2, p. 116.

50. Everett Mendelsohn, "The Ethical Implications of Western Technology for Third World Communities" (Address delivered at the Massachusetts Institute of Technology Seminar on Technology and Culture, Cambridge, Mass., 19 November 1974). Italics mine. For a revealing portrait of the differences to which Mendelsohn alludes, see *China in the Sixteenth Century: The Journals of Matteo Ricci, 1583-1610* (New York: Random House, 1953). Also Vincent Cronin, *The Wise Man from the West* (Garden City, N.Y.: Doubleday, 1957).

51. Cf. Mirrit Boutros Ghali, *Tradition for the Future* (Oxford: The Alden Press, 1972).

52. Lewis Mumford, *The Myth of the Machine*, 2 vols. (New York: Harcourt Brace Jovanovich, 1970), vol. 2, *The Pentagon of Power*, foreword.

53. On these themes, see Jeanne Hersch, "Comments on 'Industrial Society and the Good Life,' " in *World Technology and Human Destiny*, ed. Raymond Aron (Ann Arbor: University of Michigan Press, 1963), pp. 195-196.

54. See Thomas Merton, "A Note: Two Faces of Prometheus," in *The Behavior of Titans* (New York: New Directions, 1961), pp. 11-23.

55. Myron B. Bloy, Jr., "Technology and Theology," in *Dialogue on Technology*, ed. Robert Theobald (Indianapolis: Bobbs-Merrill, 1967), p. 89.

56. Cf. Arend T. van Leeuwen, *Prophecy in a Technocratic Era* (New York: Scribner's, 1968).

57. Edgar Morin, *Le paradigme perdu: la nature humaine* (Paris: Editions du Seuil, 1970), p. 229. Translation mine.

58. Ibid., p. 199.

59. Rajni Kothari, *Footsteps into the Future* (New York: The Free Press, 1974), p. 9.

60. See John and Magda C. McHale, *Human Requirements, Supply Levels and Outer Bounds: A Framework for Thinking About the Planetary Bargain* (Policy paper of the Aspen [Colo.] Institute for Humanistic Studies, Program in International Affairs, 1975); also, *The Planetary Bargain: Proposals for a New International Economic Order to Meet Human Needs* (Report of an International Workshop, Aspen Institute for Humanistic Studies, 7 July-1 August 1975).

61. Everett Reimer, "Alternative Futures for the World," unpublished book manuscript (n.p., June 1973), p. 2.

62. Mumford, *The Pentagon of Power,* pp. 10-11.

63. Examples are Robert L. Heilbroner, *An Inquiry into the Human Prospect* (New York: Norton, 1974); Barrington Moore, Jr., *Reflections on Causes of Human Misery* (Boston: Beacon, 1972); and Peter Berger, *Pyramids of Sacrifice* (New York: Basic Books, 1974).

64. Jean-Marie Domenach, *Le rétour du tragique* (Paris: Editions du Seuil, 1967).

65. Ibid., p. 13.

Index

About the Author and the Overseas Development Council

Denis Goulet is a pioneer in a new discipline, the ethics of development. His approach to field research is to experience conditions of underdevelopment in a mode of vulnerability to the adverse forces (economic, social, political) that burden the poor—who are the majorities in the particular settings in which Goulet has lived and worked in Africa, the Middle East, Europe, and Latin America. His theoretical work covers three areas: the critical examination of value conflicts and normative issues posed by competing visions of development; innovative probes in multidisciplinary research and planning methods; and development pedagogies adapted to diverse cultural and institutional contents.

Goulet holds master's degrees in philosophy and social planning and a Ph.D. in political science from the University of São Paulo. He has taught at Indiana University, the University of California (San Diego), the University of Saskatchewan, the University of Recife (Brazil), and at the Institut de Recherche et de Formation en Vue du Dêveloppement in Paris. His previous books are: *Ethics of Development* (published only in Spanish and Portuguese); *The Cruel Choice: A New Concept in the Theory of Development; The Myth of Aid: The Hidden Agenda of the Development Reports* (with Michael Hudson); and *A New Moral Order: Development Ethics and Liberation Theology.* He is at present a senior fellow at the Overseas Development Council in Washington, D.C.

The Overseas Development Council (ODC) is an independent, nonprofit organization established in 1969 to increase American understanding of the economic and social problems confronting the developing countries and of the importance of these countries to the United States in an increasingly interdependent world. The ODC seeks to promote consideration of development issues by the American public, policy-makers, specialists, educators, and the media through its research, conferences, publications, and liaison with U.S. mass membership organizations interested in U.S. relations with the developing world. The ODC's program is funded by foundations, corporations and private individuals; its policies are determined by its board of directors. The Council's president is James P. Grant. The views expressed by ODC staff members in their published studies are their own and do not necessarily represent those of the ODC, its directors, officers, or staff.

ODC Board of Directors